T0293821

Mitigation Strategies for Climate Change

Mitigation Strategies for Climate Change

Edited by Andrew Hyman

New York

Published by Syrawood Publishing House,
750 Third Avenue, 9th Floor,
New York, NY 10017, USA
www.syrawoodpublishinghouse.com

Mitigation Strategies for Climate Change
Edited by Andrew Hyman

International Standard Book Number: 978-1-64740-343-0 (Hardback)

Cataloging-in-publication Data

Mitigation strategies for climate change / edited by Andrew Hyman.
p. cm.
Includes bibliographical references and index.
ISBN 978-1-64740-343-0
1. Climatic changes. 2. Climate change mitigation. 3. Climatology. I. Hyman, Andrew.
QC903 .C55 2023
363.738 74--dc23

Table of Contents

PREFACE

This book was inspired by the evolution of our times; to answer the curiosity of inquisitive minds. Many developments have occurred across the globe in the recent past which has transformed the progress in the field.

The changes in weather patterns which are spread over a long period of time are known as climate change. Mitigation strategies for climate change refer to the actions taken to limit the human contribution in the global greenhouse effect and the accompanying climate change. Human activities are causing a rise in the concentrations of greenhouse gases globally, resulting in global warming. This effect is exacerbated by numerous self-reinforcing cycles in the Earth's system, such as the melting of sea ice increases the ocean water, which then absorbs more heat resulting in the loss of more sea ice. Furthermore, increased carbon dioxide absorption by the oceans is causing increased level of ocean acidity, which has a negative impact on marine ecosystems. Strategies for reducing greenhouse gas emissions in the atmosphere include shifting to low-carbon energy sources such as geothermal, wind power, solar, hydroelectric or nuclear. This book unravels the recent studies on the mitigation strategies being used for climate change. It will serve as a valuable source of reference for graduate and postgraduate students.

This book was developed from a mere concept to drafts to chapters and finally compiled together as a complete text to benefit the readers across all nations. To ensure the quality of the content we instilled two significant steps in our procedure. The first was to appoint an editorial team that would verify the data and statistics provided in the book and also select the most appropriate and valuable contributions from the plentiful contributions we received from authors worldwide. The next step was to appoint an expert of the topic as the Editor-in-Chief, who would head the project and finally make the necessary amendments and modifications to make the text reader-friendly. I was then commissioned to examine all the material to present the topics in the most comprehensible and productive format.

I would like to take this opportunity to thank all the contributing authors who were supportive enough to contribute their time and knowledge to this project. I also wish to convey my regards to my family who have been extremely supportive during the entire project.

Editor

1

Climate Change Adaptation Options in Farming Communities of Selected Nigerian Ecological Zones

Ayansina Ayanlade, Isaac Ayo Oluwatimilehin, Adeola A. Oladimeji, Godwin Atai and Damilola T. Agbalajobi

Contents

Abstract

This chapter examines the impacts of climate change on three tropical crops and assesses the climate change adaptation options adopted by rural farmers in the region. The study was conducted among farming communities settled in three major ecological zones in Nigeria. Over 37 years of data on rainfall and temperature were analyzed to examine climate change impacts on three major crops: rice, maize, and cassava. Farmers' adaptive capacity was assessed with a survey. Climatic data, crop yields, and survey data were analyzed using both descriptive and inferential statistics. The relation between rainfall/temperature and crop

A. Ayanlade (✉) · I. A. Oluwatimilehin · G. Atai
Department of Geography, Obafemi Awolowo University, Ile-Ife, Nigeria
e-mail: aayanlade@oauife.edu.ng; sinaayanlade@yahoo.co.uk

A. A. Oladimeji
Department of Microbiology, University of Ibadan, Ibadan, Nigeria

D. T. Agbalajobi
Department of Political Science, Obafemi Awolowo University, Ile-Ife, Nigeria

yields was examined using the Pearson correlation coefficient. Results show a high variation in the annual rainfall and temperature during the study period. The major findings from this research is that crops in different ecological zones respond differently to climate variation. The result revealed that there is a very strong relationship between precipitation and the yield of rice and cassava at $p < 0.05$ level of significance. The results further showed low level of adaption among the rural farmers. The study concludes that rainfall and temperature variability has a significant impact on crop yield in the study area, but that the adaptive capacity of most farmers to these impacts is low. There is a need for enhancing the adaptation options available to farmers in the region, which should be the focus of government policies.

Keywords

Adaptation · Climate change · Crop yield · Impacts · Nigeria

Introduction

Given their impacts on both natural and human systems, climate variability and climate change have become topical issues in recent research. As the changing nature of weather and climate directly relates to crop yields, climate change impacts on the agricultural sector and food systems have been the focus of a growing number of works, including research in many African countries (Adenle et al. 2017; Morton 2017; Serdeczny et al. 2017; Molua 2020). Such studies have reported that rainfall and minimum and maximum temperatures are the most important climatic elements for agricultural production. The scenario over the years is that these climate parameters, especially rainfall patterns, have varied, with some high confidence of change in their patterns in recent years, as perceived through indigenous knowledge of people globally (Adejuwon 2005; Ayanlade 2009; Sowunmi and Akintola 2010; Moylan 2012; Adamgbe and Ujoh 2013; Pablo and Antonio 2015; Ayanlade et al. 2017). The multi-hazard events climate change occasions comprising of windstorms, wildfire, rainstorms, droughts, and dust storms which maybe become more frequent with increased severity in the nearest future. There are several pieces of evidence in the literature that land use and land cover change (Ayanlade 2017; Fourcade et al. 2019), sea-level rise (Agboola and Ayanlade 2016; Varela et al. 2019), and changes in onset will pose significant long-term challenges. There have been changes that can be identified by alteration in the mean or variability of climate properties and that have persisted for an extended period. Consequently, climate change will pose imperative short-term and long-term bottlenecks to agricultural production, particularly for rural farmers who depend on rainfall for cultivation (IPCC 2007; Heltberg et al. 2009; Ayanlade et al. 2018b; Ayanlade and Ojebisi 2020).

Agriculture is vulnerable to climate change as crops are sensitive to rising temperatures and changing rainfall patterns. Rising temperatures and erratic rainfalls threat food crop production, thus exacerbating food insecurity and poverty (Sanchez 2000;

Oluoko-Odingo 2011; Seaman et al. 2014). The impacts of climate change on agriculture are likely to intensify in the future as climate models have predicted increasing temperature and more erratic rainfall with the potential increase in the intensity and frequency of extreme weather events (Wright et al. 2014; Ayanlade et al. 2018a; Hein et al. 2019). For example, some studies have suggested that climate change will reduce the yield of crops like maize, rice, and cassava by 15–25%, partly because climate change will alter the incidence and severity of pest and disease outbreaks, indirectly affecting crop production and yield (Harvey et al. 2018). Even though climate change portends serious dangers to agriculture, current studies on climate change impacts on the agricultural systems of Nigeria suffer from several gaps. First, previous studies have mostly focused on crops like yam, coffee, cocoa, sorghum, etc. (Ayanlade 2009; Ayanlade et al. 2009; Ajetomobi 2016). Other studies on the topic have used short-term data, although long-term data is more accurate in delineating climate change effects on agricultural practices (Adejuwon 2005, 2012). Thus, previous research on this topic has treated Nigeria as a single ecological zone. To fill these gaps, the present study focuses on climate change impacts on three crops that have not previously received attention (i.e., rice, maize, and cassava) and using data from three (3) different states in Nigeria, falling within three different ecological zones: Ondo (Rainforest), Ogun (Freshwater), and Kwara state (Guinea savanna). This spread will importantly give room for spatiotemporal comparison. More so, this study involves the indigenous knowledge of climate change, effects, impacts, and adaptation practices among the rural farmers. Rural farming is very important in this part of the world for the provision of food and productive employment for many people.

Climate Change Assessment in Selected Ecological Zones in Nigeria

Research as conducted in three different sites (Fig. 1), representing three different ecological zones: Ondo (rainforest located within latitude 7°10′N, 7°15′N and longitudes 5°5′E and 5°83′E); Ogun (freshwater, located between on latitude 6°12′N and 7°47′N and longitude 3°0′E and 5°0′E); and Kwara (savanna, located between latitude 7°45′N and 9°30′N and longitude 2°30′ and 6°25′E). The three sites fall within the tropical wet and dry climate of Koppen's climate classification, dominated by a wet and a dry season. There are, however, some variations between sites. The climate of Ondo State corresponds to a lowland tropical rain forest type, with a mean monthly temperature ranging from 19 °C to 30 °C (mean monthly value of 27 °C) and a mean relative humidity 75% (Owoeye and Sekumade 2016). The mean annual total rainfall exceeds 2,000 mm. In the northern part of the state, there is marked dry season from November to March with little or no rain, for which the total annual rainfall in the north is of about 1,800 mm (Jamie 2016). Differently, Ogun state is characterized by a tropical climate consisting of two distinct wet and dry seasons. The long dry season extends from November to March. The annual rainfall value ranges between 1,400 and 1,500 mm with a relatively high temperature of an average of 30 °C. The average temperature value varies from one month to another,

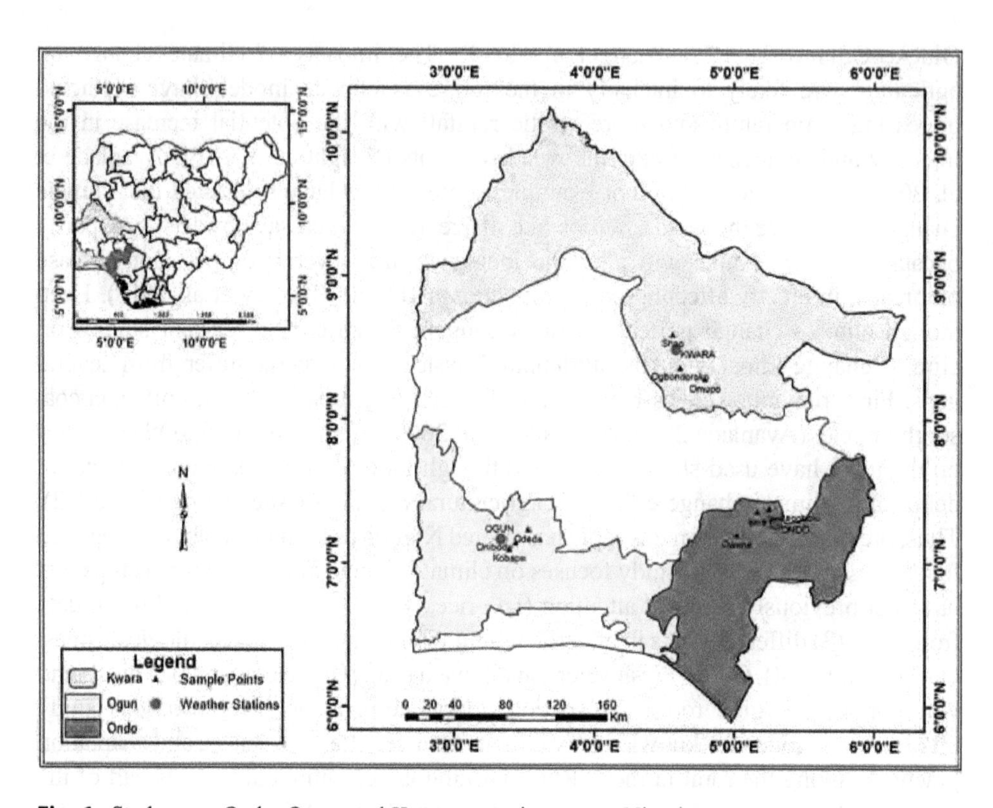

Fig. 1 Study area: Ondo, Ogun, and Kwara states in western Nigeria

with a minimum average of 25.7 °C in July and a maximum of 30.2 °C in February. The humidity is lowest at the peak of the dry season in February usually at 37–54% and highest between June and September with a value of 78–85% (Adeleke et al. 2015). Kwara state (Oladimeji et al. 2015) enjoys an annual rainfall ranging from 1,000 mm to 1,500 mm with the rainy season beginning at the end of March and lasting until September while the dry season starts in October and ends in March (Oriola et al. 2010). The temperature of the state is uniformly high and ranges between 25 °C and 30 °C throughout the wet season except for July and September in which the cloud cover prevents direct insolation. In the dry season, the temperature ranges between 33 °C and 34 °C. In the rainy season, the relative humidity in Kwara state ranges between 75% and 80% while in the dry season it is about 65% (Akpenpuun and Busari 2013).

Ondo is an agrarian state with large-scale production of crops like cocoa, yam, cocoa, coffee, and rubber and huge forest reserves of about 2,008 km^2 which produce timbers for furniture, fuelwood, and industrial uses. Ogun is also an agrarian state with the production of arable crops like maize, rice, cassava, and melon. Several other identifiable modern economic activities exist in Ogun, including insurance, motor companies, petrol stations, and light and heavy industries. The major economic activity of Kwara state is farming, with a particular predominance of cash crops like cocoa, coffee, kolanut, tobacco, beniseed, and palm. Food crops like cereals (rice, maize, millet) and tuber crops (yam, cassava, cocoyam, etc.) are grown in Kwara (Oluwasusi and Tijani 2013).

Daily climatic data of rainfall and minimum and maximum temperatures were used for this study. The data, spanning for a period of 37 years (1982–2017), were collected from the synoptic stations (Fig. 1) located at Akure (Ondo state), Ilorin (Kwara state), and Abeokuta (Ogun state). The data were collected by the Nigerian Meteorological Agency (NIMET) at the stations using the British Standard Rain gauge and Dine's tilting siphon rainfall recorder for rainfall and thermometer for temperature. To establish the pattern of climate, a year running mean was used in the analysis to show the annual fluctuation of rainfall and temperature. Mean of daily data were averaged to get the annual mean for temperature (minimum and maximum) and rainfall using MS Excel. Sigma Plot (Version 10.2) was used to plot the graph for annual rainfall, and minimum and maximum temperature against the year covering the study period; the curves were then fitted to show the linear trend. Three crops, that is, maize (cereal), rice (cereal), and cassava (tuber), were used for this study. In this research, climate factors are seen as one of the important requirement for the yield of the crops, as normal seasonal water requirements of maize, for example, is between 400 and 600 mm (Phiri et al. 2003). Maize and rice are common crops in all three zones. Crop yield data were obtained from the archives of Agricultural Development Project (ADP) offices in Ondo, Ogun, and Kwara states. Data were analyzed using the Pearson correlation coefficient between crop yields and climatic data. This was done using the formula below:

$$r = \frac{N\sum xy - (\sum x)(\sum y)}{\sqrt{\left[\sum x^2 - (\sum x)^2\right]\left[N\sum y^2 - (\sum y)^2\right]}}$$

where N is the number of pairs of scores, $\sum xy$ is sum of the products of paired scores, $\sum x$ is sum of scores, $\sum y$ is sum of y scores, $\sum x^2$ is sum of squared x scores, $\sum y^2$ is sum of squared y scores, x is the crop yield, and y is the climate data.

Social survey data for the study were collected from farmers with the aid of questionnaires and focus group discussion. Our survey instrument was designed to assess farmer's adaptive capacity. Data were collected among farmers who are purposively selected across the three states under consideration. Criteria for inclusion in our sample were to be in a settlement where maize, rice, and cassava are cultivated and to be 30 years and above. These criteria ensured the collection of data among experienced farmers who had witnessed different farming seasons. The responses from the questionnaires were analyzed using both descriptive and inferential statistics. The major aim of social survey is to obtain information from the farmers on how the yields have been over the years and compare it with changes in climate.

Annual Variability in Rainfall and Minimum and Maximum Temperature

Rainfall data for a 37 years period revealed a variation in the annual mean amount of rainfall (Fig. 2). Rather than an absolute change, results for all the three stations show rainfall variability. The rainfall in Kwara is below normal, a little above normal

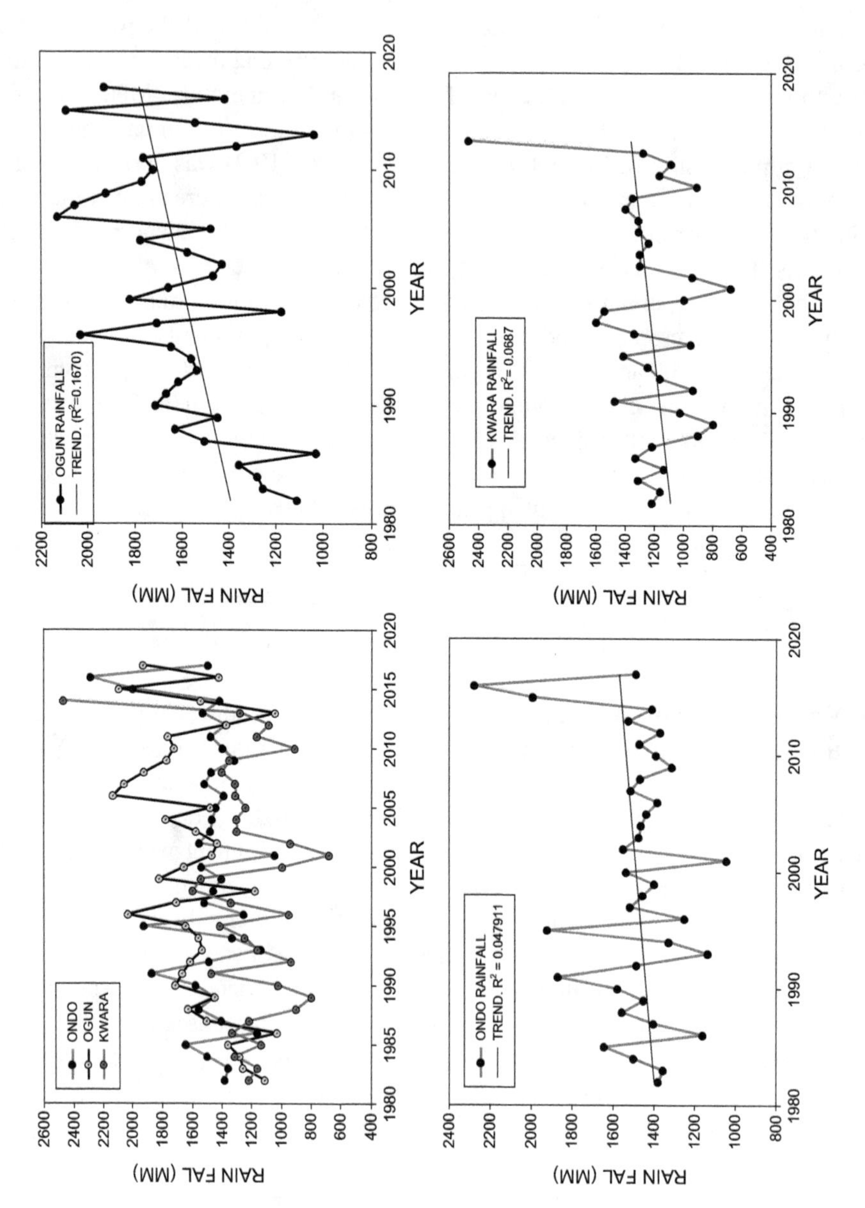

Fig. 2 Rainfall variability in the study area

in Ogun, and varied in Ondo. The rainfall variability found in the data can be attributed to seasonal and interannual climatic variability, which is much more several particularly in the savannah region where Kwara is located (Adejuwon 2004; Odekunle et al. 2007; Ayanlade 2009). As depicted in Fig. 2, annual rainfall was mostly below average before 2010, but varies greatly in the last decade, with readings above average in all the stations. Rainfall peaked in 1991 (with 1,982 mm) and 1995 (1,999 mm) in Ondo state, with 2016 (2,206 mm) recording the highest rainfall during the study period (Fig. 2). In Ogun state, rainfall peaked in 1996 (2,026 mm), 2006 (2,216 mm), and 2015 (2,086 mm), with 2006 recording the highest rainfall. Finally, in Kwara, rainfall peaked in 1991 (1,468 mm), 1995 (1,409 mm), 1998 (1,596 mm), and 1999 (1,539 mm), with the highest peak recorded in 2014 (2,467 mm) with a very unusual high value. Generally, the annual average of rainfall was lowest (673 mm) in Kwara state, with similar values in Ogun (1,028 mm in 1986 and 1,037 mm in 2013) and Ondo states (1,041 mm). Values follow an expected pattern, with higher values in the rainforest ecological zone close to the ocean and lower values in inward zones closer to the savanna. Ogun state received the highest rainfall between 1995 and 2010. The highest rainfall within a year was recorded in 2014 in Kwara, the state belonging to Guinea savanna ecological zone, which recorded a value slightly above the peak of rainfall in Ondo (which belongs to the forest ecological zone).

Climatic variability in the study area can be linked to global climate oscillating systems as El Niño – Southern Oscillations, sea surface temperatures (SSTs), and Inter-Tropical Discontinuity (ITD) have been reported to be responsible for inter-annual variability in Africa climate (Stige et al. 2006). Such climate variability can cause negative departures from normal climate, as documented in a part of Nigeria and elsewhere (Ashipala 2013; Owusu et al. 2015; Ayanlade et al. 2018a). Studies have further revealed that rainfall associated with SST strongly influences rainfall variability in Nigeria in conjunction with the role of the ITD as the equatorial displacement of the Atlantic Subtropical High suppresses the northward summer migration of the ITD thereby resulting into rainfall variability (Bello 2008; Ayanlade et al. 2019).

High variability was observed in the maximum temperature of all stations between 1982 and 2017 (Fig. 4). The temperature was highest in Kwara state, the northernmost state which falls within the Guinea savanna – an area dominated with relatively low rainfall and high temperature – and lowest in Ondo state, the southernmost state. Ondo and Ogun state recorded temperatures below average between 1982 and 2010. However, temperatures have experienced an increase from 2012 till date, especially in Ogun state with the temperature higher gradually increases. As shown in Fig. 3, all the stations experienced similar minimum temperatures patterns from the early 1980s. Temperature oscillations in the maximum temperature were recorded between 1990 and 2010 in Ondo and Ogun with an upward trend observed only in Kwara, though from 2010, the temperature has generally increased in all stations (Figs. 3 and 4).

As shown in Fig. 3, the annual minimum temperature is highest in Ogun state with the lowest value in 1994 (22.8 °C), a value that coincides with the highest

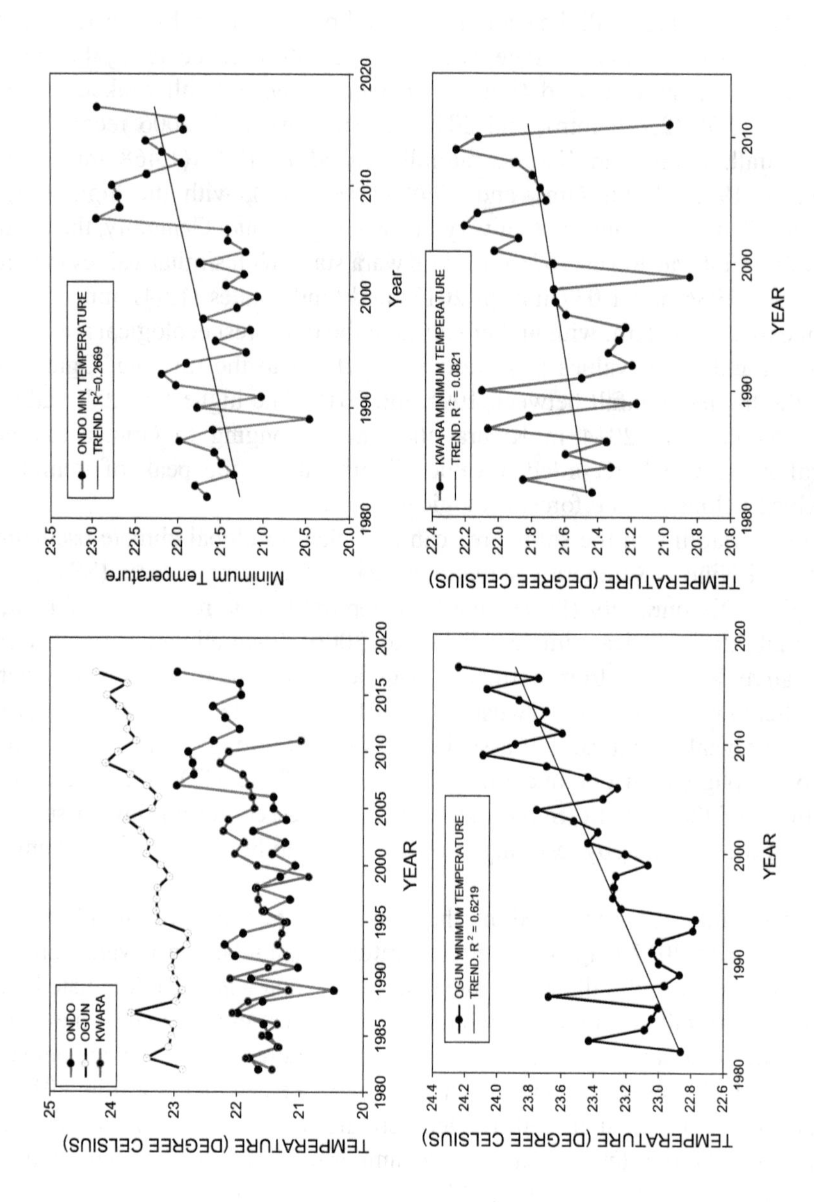

Fig. 3 Minimum temperature variability

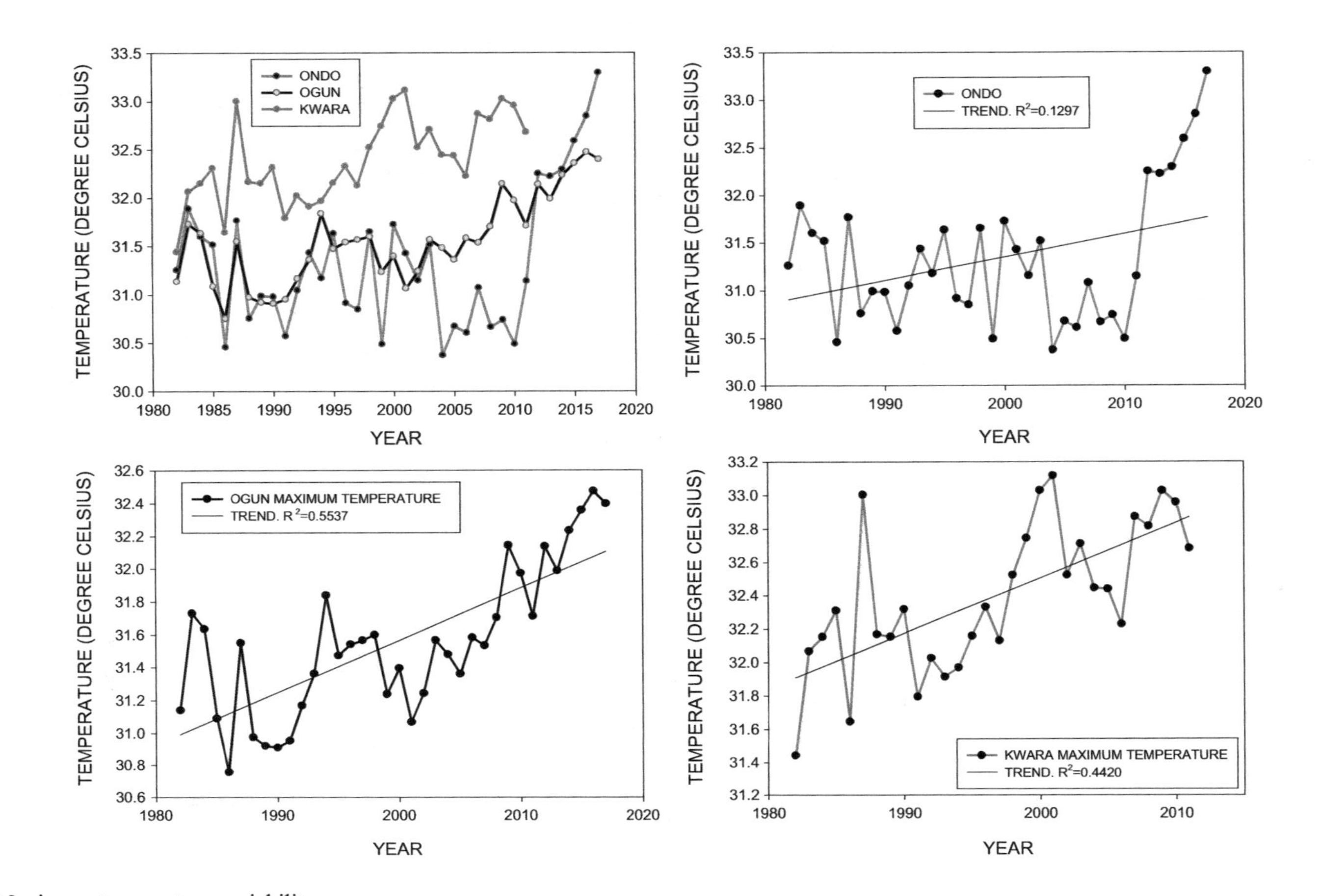

Fig. 4 Maximum temperature variability

minimum temperature in Ondo state. Despite the southern location of Ogun state, the minimum temperature is generally above the minimum temperature for the other two stations, among which Ondo displays the lowest temperature than Kwara. These values represent the highest temperature difference in Kwara and the lowest temperature difference in Ogun state. The period of temperature increase in Kwara state (i.e., years 2000, 2001, and 2002) corresponds to the period of rainfall decrease which could be attributed to increasing temperature and decreasing rainfall as a result of the increasing evapotranspiration and desertification also reported in other areas of Nigeria (Ayoade 2003). The trend of the maximum temperature varies considerably in all stations. However, while the trend is irregular in the early years of the study period, an upward increasing trend was observed in the later years. This recent increase in temperature could be linked to the effect of changes in climate and to the rapid urbanization in the ecological zones under consideration (Mabo 2006).

Relationship Between Climate and Crop Yield

The results of the Pearson correlation analysis between annual minimum temperature and crop yields during the growing season are shown in Table 1. Annual minimum temperature had positive but weak correlation with maize in Ondo (0.142) but not significant at $p < 0.05$, a weak and not statistically significant negative correlation of −0.078 at $p < 0.05$ in Ogun state, and a positive correlation of 0.219 in Kwara at $p < 0.05$. Rice, on the other hand, had a positive but weak correlation of 0.167 which is statistically not significant at $p < 0.05$ in Ondo state while in Ogun state with a positively strong and statistically significant correlation coefficient of 0.674 at $p < 0.05$ and a strong positive and statistically significant correlation coefficient of 0.481 in Kwara state at $p < 0.05$. The correlation coefficient for cassava in Ondo state is a weak and statistically not significant value of 0.22 at $p < 0.05$ while a strong positive and statistically significant correlation coefficient of 0.82 at $p < 0.05$ in Ogun and a positive but not significant correlation of 0.25 at $p < 0.05$. Generally, annual minimum temperature had a strong, positive, and statistically significant correlation coefficient at $p < 0.01$ and $p < 0.05$ in rice for Ogun and Kwara states and cassava only in Ogun state.

Table 2 shows the result of Pearson correlation analysis between annual maximum temperature and crop yields during the study period. The annual maximum temperature had a positive weak and statistically not the significant relationship with maize in Ondo (0.198) at $p < 0.05$, a weak and not statistically significant negative

Table 1 Correlation between minimum temperature and crop yield

	Maize	Rice	Cassava
Ondo	0.142	0.167	0.220
Ogun	−0.078	0.674*	0.820*
Kwara	0.219	0.481*	0.248

*Significant at $p < 0.05$

Table 2 Correlation between maximum temperature and crop yield

	Maize	Rice	Cassava
Ondo	0.198	0.160	−0.115
Ogun	−0.337	0.884**	0.870**
Kwara	0.342	0.773**	0.626**

*Significant at $P \leq 0.05$; **Significant at $P \leq 0.01$

Table 3 Correlation between rainfall and crop yield

	Maize	Rice	Cassava
Ondo	−0.151	−0.091	−0.076
Ogun	0.003	0.136	0.046
Kwara	0.08	−0.220	0.064

*Significant at $P \leq 0.05$; **Significant at $P \leq 0.01$

correlation of −0.337 at $p < 0.05$ in Ogun state, and a positive correlation of 0.342 in Kwara at $p < 0.05$. Rice had a positive but weak correlation of 0.160 which is statistically not significant at $p < 0.05$ in Ondo state while in Ogun state with a positive, very strong, and statistically significant correlation coefficient of 0.88 at $p < 0.05$ and a very strong positive and statistically significant correlation coefficient of 0.77 in Kwara state at $p < 0.05$. The correlation coefficient for cassava in Ondo state is a negative, weak, and statistically not significant value of −0.12 at $p < 0.05$ while a very strong positive and statistically significant correlation coefficient of 0.87 at $p < 0.05$ in Ogun. A positive and significant correlation of 0.63 at $p < 0.05$. Generally, the annual maximum temperature had a strong, positive, and statistically significant association with rice and cassava yields in Ogun and Kwara states but not in Ondo state. The maximum temperature was not associated with maize yields in any of the states.

The result of the Pearson correlation analysis between annual rainfall and crop yields during the study period is shown in Table 3. Annual rainfall had a negative, weak, and statistically not a significant relationship of −0.15 with maize Ondo at $p < 0.05$, a weak and not statistically significant positive correlation of 0.003 at $p < 0.05$ in Ogun state and a positive correlation of 0.08 in Kwara at $p < 0.05$. Rice showed a negative but weak correlation of −0.091 which is statistically not significant at $p < 0.05$ in Ondo state while in Ogun state with a positive, weak, and statistically significant correlation coefficient of 0.14 at $p < 0.05$ and a weak, negative, and statistically not significant correlation coefficient of −0.22 in Kwara state at $p < 0.05$. The correlation coefficient for cassava in Ondo state is a negative, weak, and statistically not significant value of −0.08 at $p < 0.05$ while a positive and statistically not significant correlation coefficient of 0.046 at $p < 0.05$ in Ogun was observed. And a positive, weak, and not the significant correlation of 0.06 at $p < 0.05$ was observed for Kwara state. Generally, annual rainfall had a weak and statistically not significant association with maize, rice, and cassava yields in the three studied states.

Farmer's Adaptive Strategies

Figure 5 shows the percentage error graph of farmers' coping or adaptive capacity toward extreme climatic conditions. According to our results, 16.7% of the respondents are engaged in agricultural diversification, 17.3% are engaged in changing the crops they cultivate, 20.2% are engaged crop rotation, and 19.6% are practicing mixed farming.

As shown in Fig. 5, the farmers are coping or adapting to extreme climatic conditions but generally with low capacity. This is because the results show that 19% of farmers had implemented changes in the size of farmland, 14.9% are engaged in different crop composition, 33.3% are engaged in agricultural intensification, 30.4% are engaged in bush fallowing, 29.8% are changing the daily working time, 47.6% are presently engaged in agroforestry, 27.4% are changing the seasonal timing of sowing, 41.1% are changing the harvest time, 35.7% are into irrigation, 31.5% are into the use of fertilizer, 32.1% are into the use of pesticide, 56.5% are into getting loan and credit facilities, 32.7% are into getting other sources of income so as not to depend on their farm only, 44% are changing the method of storing their food, 27.4% are changing the quantity of food consumed by the family, and 24.4%

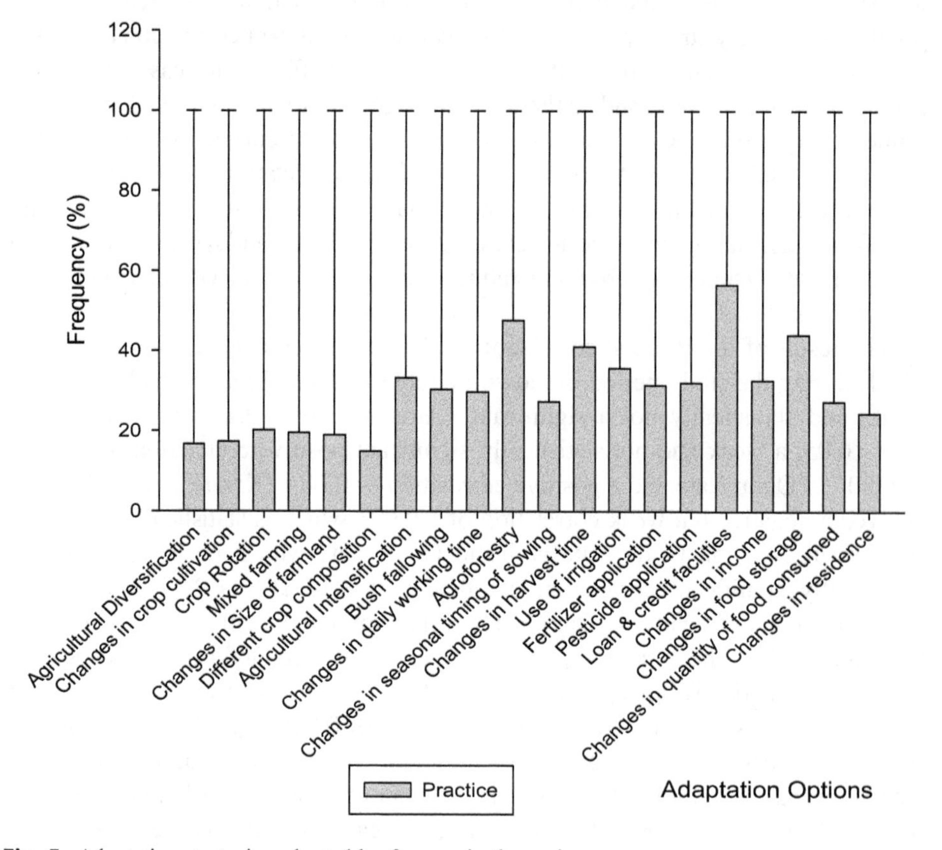

Fig. 5 Adaptation strategies adopted by farmers in the region

are changing their residence by moving to another area for farming. From Fig. 5, seeking for loans and credit facilities, changes in food storage, agroforestry, changing the time of harvest, use of irrigation, use of fertilizer, and agricultural intensification, among others, are the most practiced adaptation options.

Conclusion

The effects of climate variability or climate change on selected crops in selected ecological zones in Nigeria were examined in this study. Results show that minimum temperature, maximum temperature, and rainfall varied both spatially and temporally. The annual minimum temperature, maximum temperature, and rainfall showed that there is the year-to-year fluctuation with pronounced peaks and depressions. While some years show upward trends, some showed a downward trend. Also, there is the month to month variability in minimum temperature, maximum temperature, and rainfall. Crop yields did not depend on rainfall and minimum and maximum temperatures alone. In all the stations, maize yields were less sensitive to minimum temperature than rice and cassava yields. While there appears a negative but weak relationship that is not statistically significant between minimum temperature and maize in Ogun, it has a positive but weak relationship in Ondo and Kwara. Maximum temperature showed a strong positive and statistical association with rice and cassava yields in Ogun and Kwara states. Finally, rainfall was associated with maize, rice, or cassava yields in any state. Many other studies have reported the similar finding of the climate-crops relationships in China (Zhang et al. 2016; Gao et al. 2019; Ding et al. 2020), Southern Africa (Mafongoya et al. 2017; Nhamo et al. 2019) and East African countries (Sridharan et al. 2019; Thomas et al. 2019). These results dovetail with results from previous research. Thus, although an increasing trend in rainfall and temperature may affect rice yields, a study in Bangladesh showed that despite an increasing rainfall and temperature, the yield of rice is not negatively affected (Rahman et al. 2017). Matched with the results of the present study, the rainfall can, therefore, have a positive or negative impact on the yield of crops depending on its distribution and intensity over the growing season. Furthermore, the present study revealed that farmers' perception of climate variability/change impacts on crop yield varied spatially. The majority of farmers believed that average temperature of the coldest month, rain generally, rain during the rainy season, and rain during the dry season were believed to have changed in recent years in Ogun and Kwara by higher numbers of respondents, while statistical analyses showed that there is variation while in Ondo state, the farmers believe that "the temperature is warmer generally, the temperature of the hottest season is warmer, rain is higher in the rainy and dry seasons with extreme drought less frequent, onset earlier, rainy season duration longer." The temperature of the hottest month was believed, though, to be warmer by a higher proportion, as claimed by the farmers. Many believe that extreme floods are more frequent while extreme drought and duration of the rainy and dry seasons have remained the same. According to Tripathi and Mishra (2017), farmers are aware of long-term changes in climatic factors (temperature and rainfall) but are unable to identify those changes as climate change. However, many farmers are not taking concrete steps in dealing with perceived change in climate but are changing their agricultural and farming practices (Ayanlade et al. 2018a) which can be said to be passive response to climate change. Farmers' perception of climate is probably responsible for low adaptation to climate change.

Acknowledgments This study is part of the research project carried out with the aid of a grant from the International Development Research Centre, Ottawa, Canada, and with financial support from the Government of Canada, provided through Global Affairs Canada (GAC), under the framework of the Mathematical Sciences for Climate Change Resilience (MS4CR) program administered by AIMS (2018 cohort of AIMS). The research was also partly funded by an ERC Consolidator Grant to Reyes-García (FP7-771056-LICCI).

References

Adamgbe E, Ujoh F (2013) Effect of variability in rainfall characteristics on maize yield in Gboko, Nigeria. J Environ Prot 4:881

Adejuwon JO (2004) Crop yield response to climate variability in the Sudano-Sahelian ecological zones of Nigeria in southwestern Nigeria. In: AIACC report of workshop for Africa and Indian Ocean Island. Dakar, Senegal pp (15–16)

Adejuwon JO (2005) Food crop production in Nigeria. I. Present effects of climate variability. Clim Res 30:53–60

Adejuwon JO (2012) An assessment of the effect of climate variability on selected agricultural practices and yields in Sokoto-Rima River Basin, Nigeria. PhD thesis, Obafemi Awolowo University, Ile-Ife

Adeleke O, Makinde V, Eruola A, Dada O, Ojo A, Aluko T (2015) Estimation of groundwater recharges in Odeda local government area, Ogun state, Nigeria using empirical formulae. Challenges 6:271–281

Adenle AA, Ford JD, Morton J, Twomlow S, Alverson K, Cattaneo A, Cervigni R, Kurukulasuriya P, Huq S, Helfgott A (2017) Managing climate change risks in Africa – a global perspective. Ecol Econ 141:190–201

Agboola AM, Ayanlade A (2016) Sea level rise and its potential impacts on coastal urban area: a case of Eti-Osa, Nigeria. An Univ Oradea Ser Geogr 27:188–200

Ajetomobi J (2016) Sensitivity of crop yield to extreme weather in Nigeria. No 310-2016-5403 2016

Akpenpuun T, Busari R (2013) Impact of climate on tuber crops yield in Kwara state, Nigeria. Am Int J Contemp Res 3:52–57

Ashipala SN (2013) Effect of climate variability on pearl millet (Pennisetum glaucum) productivity and the applicability of combined drought index for monitoring drought in namibia. A Dissertation submitted for the award of M.Sc Agricultural Meteorolgy, University of Nairobi, Nairobi, Kenya

Ayanlade A (2009) Seasonal rainfall variability in Guinea Savanna part of Nigeria: a GIS approach. Int J Clim Change Strateg Manage 1:282–296

Ayanlade A (2017) Variations in urban surface temperature: an assessment of land use change impacts over Lagos metropolis. Weather 72:315–319

Ayanlade A, Ojebisi SM (2020) Climate variability and change in Guinea Savannah ecological zone, Nigeria: assessment of cattle herders' responses. In: Leal Filho W (ed) Handbook of climate change resilience. Springer International Publishing, Cham

Ayanlade A, Odekunle T, Orimogunje O, Adeoye N (2009) Inter-annual climate variability and crop yields anomalies in Middle Belt of Nigeria. Adv Nat Appl Sci 3:452–465
Ayanlade A, Radeny M, Morton JF (2017) Comparing smallholder farmers' perception of climate change with meteorological data: a case study from southwestern Nigeria. Weather Clim Extrem 15:24–33
Ayanlade A, Radeny M, Akin-Onigbinde AI (2018a) Climate variability/change and attitude to adaptation technologies: a pilot study among selected rural farmers' communities in Nigeria. GeoJournal 83:319–331
Ayanlade A, Radeny M, Morton JF, Muchaba T (2018b) Rainfall variability and drought characteristics in two agro-climatic zones: an assessment of climate change challenges in Africa. Sci Total Environ 630:728–737
Ayanlade A, Atai G, Jegede MO (2019) Variability in atmospheric aerosols and effects of humidity, wind and intertropical discontinuity over different ecological zones in Nigeria. Atmos Environ 201:369–380
Ayoade J (2003) Climate change: a synopsis of its nature, causes, effects and management. Vantage Publishers, Ibadan
Bello NJ (2008) Perturbations in the plant environment: the threats and agro-climatological implications for food security. University of Agriculture, Abeokuta
Ding Y, Wang W, Zhuang Q, Luo Y (2020) Adaptation of paddy rice in China to climate change: the effects of shifting sowing date on yield and irrigation water requirement. Agric Water Manag 228:105890
Fourcade Y, Åström S, Öckinger E (2019) Climate and land-cover change alter bumblebee species richness and community composition in subalpine areas. Biodivers Conserv 28:639–653
Gao J, Yang X, Zheng B, Liu Z, Zhao J, Sun S, Li K, Dong C (2019) Effects of climate change on the extension of the potential double cropping region and crop water requirements in Northern China. Agric For Meteorol 268:146–155
Harvey CA, Saborio-Rodríguez M, Martinez-Rodríguez MR, Viguera B, Chain-Guadarrama A, Vignola R, Alpizar F (2018) Climate change impacts and adaptation among smallholder farmers in Central America. Agric Food Secur 7(1):57
Hein Y, Vijitsrikamol K, Attavanich W, Janekarnkij P (2019) Do farmers perceive the trends of local climate variability accurately? An analysis of farmers' perceptions and meteorological data in Myanmar. Climate 7:64
Heltberg R, Siegel PB, Jorgensen SL (2009) Addressing human vulnerability to climate change: toward a 'no-regrets' approach. Glob Environ Chang 19:89–99
IPCC (2007) Climate Change 2007 – impacts, adaptation and vulnerability. Contribution of working group ii to the fourth assessment report of the IPCC (eds: Parry ML, Canziani OF, Palukitof JP, Van der Linden PJ, Hanson CE). Cambridge University Press, Cambridge
Jamie T (2016) Ondo state. https://en.wikipedia.org/wiki/Ondo_State. Accessed 24 Oct 2018
Mabo C (2006) Temperature variation in northern Nigeria between 1970 and 2000. J Energy Environ 19:80–88
Mafongoya PL, Peerbhay K, Jiri O, Nhamo N (2017) Climate scenarios in relation to agricultural patterns of major crops in Southern Africa. In: Smart technologies for sustainable smallholder agriculture (pp 21–37). Academic Press
Molua EL (2020) Africa in a changing climate: redefining Africa's agrarian development policies. In: Climate variability and change in Africa (pp 171–181). Springer, Cham
Morton J (2017) Climate change and African agriculture: unlocking the potential of research and advisory services. In: Nunan F (ed) Making climate compatible development happen (pp 109–135). Routledge
Moylan H (2012) The impact of rainfall variability on agricultural production and household welfare in rural Malawi. In: Uo I (ed) Urbana-Champaign. University of Illinois, Illinois
Nhamo L, Matchaya G, Mabhaudhi T, Nhlengethwa S, Nhemachena C, Mpandeli S (2019) Cereal production trends under climate change: impacts and adaptation strategies in southern Africa. Agriculture 9:30
Odekunle T, Orinmoogunje I, Ayanlade A (2007) Application of GIS to assess rainfall variability impacts on crop yield in Guinean Savanna part of Nigeria. Afr J Biotechnol 6:2100

Oladimeji Y, Abdulsalam Z, Abdullahi A (2015) Determinants of participation of rural farm households in non-farm activities in Kwara state, Nigeria: a paradigm of poverty alleviation. Ethiop J Environ Stud Manag 6:635–649

Oluoko-Odingo AA (2011) Vulnerability and adaptation to food insecurity and poverty in Kenya. Ann Assoc Am Geogr 101:1–20

Oluwasusi J, Tijani S (2013) Farmers adaptation strategies to the effect of climate variation on yam production: a case study in Ekiti state, Nigeria. Agrosearch 13:20–31

Oriola E, Ifabiyi I, Hammed A (2010) Impact of reforestation in a part of degrading natural ecological system of Ilorin, Kwara state, Nigeria. Afr J Agric Res 5:2811–2816

Owoeye R, Sekumade A (2016) Effect of climate change on cocoa production in Ondo state, Nigeria. J Soc Sci Res 10:2014–2025

Owusu K, Obour PB, Asare-Baffour S (2015) Climate variability and climate change impacts on smallholder farmers in the Akuapem North District, Ghana. In: Handbook of climate change adaptation (pp 1791–1806). Berlin: Springer

Pablo M, Antonio Y (2015) The effect of rainfall variation on agricultural households. In: International conference of agricultural economists, Milan

Phiri E, Verplancke H, Kwesiga F, Mafongoya P (2003) Water balance and maize yield following improved sesbania fallow in eastern Zambia. Agrofor Syst 59:197–205

Rahman MA, Kang S, Nagabhatla N, Macnee R (2017) Impacts of temperature and rainfall variation on rice productivity in major ecosystems of Bangladesh. Agric Food Secur 6:10

Sanchez PA (2000) Linking climate change research with food security and poverty reduction in the tropics. Agric Ecosyst Environ 82:371–383

Seaman JA, Sawdon GE, Acidri J, Petty C (2014) The household economy approach. Managing the impact of climate change on poverty and food security in developing countries. Clim Risk Manag 4:59–68

Serdeczny O, Adams S, Baarsch F, Coumou D, Robinson A, Hare W, Schaeffer M, Perrette M, Reinhardt J (2017) Climate change impacts in sub-Saharan Africa: from physical changes to their social repercussions. Reg Environ Chang 17:1585–1600

Sowunmi F, Akintola J (2010) Effect of climatic variability on maize production in Nigeria. Res J Environ Earth Sci 2:19–30

Sridharan V, Pereira Ramos E, Zepeda E, Boehlert B, Shivakumar A, Taliotis C, Howells M (2019) The impact of climate change on crop production in Uganda – an integrated systems assessment with water and energy implications. Water 11:1805

Stige LC, Stave J, Chan KS, Ciannelli L, Pettorelli N, Glantz M, Stenseth NC (2006) The effect of climate variation on agro-pastoral production in Africa. Proceedings of the National Academy of Sciences 103(9):3049–3053

Thomas TS, Dorosh PA, Robertson RD (2019) Climate change impacts on crop yields in Ethiopia, vol 130. International Food Policy Research Institute, Washington, DC

Tripathi A, Mishra AK (2017) Knowledge and passive adaptation to climate change: an example from Indian farmers. Clim Risk Manag 16:195–207

Varela MR, Patrício AR, Anderson K, Broderick AC, DeBell L, Hawkes LA, Tilley D, Snape RT, Westoby MJ, Godley BJ (2019) Assessing climate change associated sea-level rise impacts on sea turtle nesting beaches using drones, photogrammetry and a novel GPS system. Glob Chang Biol 25:753–762

Wright H, Vermeulen S, Laganda G, Olupot M, Ampaire E, Jat M (2014) Farmers, food and climate change: ensuring community-based adaptation is mainstreamed into agricultural programmes. Clim Dev 6:318–328

Zhang Z, Song X, Tao F, Zhang S, Shi W (2016) Climate trends and crop production in China at county scale, 1980 to 2008. Theor Appl Climatol 123:291–302

2

Agro-Ecological Lower Midland Zones IV and V in Kenya using GIS and Remote Sensing for Climate-Smart Crop Management

Hilda Manzi and Joseph P. Gweyi-Onyango

Contents

Abstract

Food production in Kenya and Africa in recent past has experienced vagaries of weather fluctuations which ultimately have affected crop yield. Farming in Kenya is localized in specific Agro-ecological zones, hence understanding crop growth responses in particular regions is crucial in planning and management for purposes of accelerating adoption. A number of strategies for adoption and adaptation to changing weather patterns have been deployed yet only limited challenges have been partially addressed or managed. This chapter examines previous methods used in classifying agro-ecological zones and further provides additional insightful parameters that can be adopted to enable farmers understand and adapt better to the current

H. Manzi (✉) · J. P. Gweyi-Onyango
Department of Agricultural Science and Technology, Kenyatta University, Nairobi, Kenya
e-mail: gweyi.joseph@ku.ac.ke

variable and unpredictable cropping seasons. The chapter scrutinizes past and current documented information on agro-ecological zonal valuations coupled with the use of earth observation components such as air temperature at surface, land surface temperature, evapotranspiration, soil, temperature, and soil and moisture content in order to better understand and effectively respond to new phenomena occurring as a result of climate change in the marginal agricultural areas. Significant variations in precipitation, ambient temperature, soil moisture content, and soil temperature become evident when earth observation data are used in evaluation of agro-ecological lower midland zones IV and V. The said variations cut across areas within the agro-ecological zones that have been allocated similar characteristics when assigning cropping seasons. The chapter summarizes the outcomes of various streams of contributions that have reported significant shifts or changes in rainfall and temperature patterns across Kenya and wider Eastern Africa region. The chapter highlights the need for re-evaluation of the agro-ecological zones based on the recent earth observation datasets in their diversity. The research emphasizes the use of multiple climate and soil-related parameters in understanding climate change in the other marginal areas of Kenya.

Keywords

Climate change · Length of growth period · Cropping seasons · Earth observation data

Introduction

Climate Change and Its Impacts in Lower Midland Agro-ecological Zones IV and V

The worrying trends and implication of the changing climatic patterns and extreme weather events have been previously documented (Donat et al. 2013; Schneider et al. 2018). One of the major concerns is the shift in weather patterns and its adverse effect on rain-fed agriculture. Previous work (Pearce 2000; Rosenzwerg, and le Parry 1994) have shown the vulnerability of most developing countries to climate change. Seo et al. (2009) evaluated the effects of climate variability in Africa and singled out the semi-arid areas as the most sensitive while the productive areas such as the high to medium wet zones becoming even more useful for agriculture. These researchers highlight the role that agro-ecological zones play to bring out the understanding of climate change and further emphasize the role of climate change on the impact on crops and livestock in light of changing economic value. They further observed that climate resilience is built more on net crop and livestock revenue combined rather than crop or livestock alone, particularly the African region. Mendelsohn (2008) links markets and places the largest economic impact of climate change upon agriculture. Furthermore, studies in developing countries have pointed to challenges in placing the economic value to agriculture due to lack of data on farm performance, as argued by Mendelsohn (2008). The author draws parallels from tropical and subtropical agriculture and is of the opinion that both regions are more sensitive to climate change compare to temperate zones and that the level of effect emanate more on climate scenarios at hand than agriculture practices or level of farming activities. In addition, the climate variability differs from one country to another and differences are also evident within regions in the countries (Mendelsohn 2008).

Kenya has experienced extreme rainfall events in every 3 years cycle on average based on the 1989–2011 analysis (Ndirangu et al. 2017). In the same period, severe droughts and changes in rainfall variability have become inevitable. These have severely affected farmers who depend on rain-fed agriculture and have eventually become victims of climate variability and extreme weather events. Furthermore, the global financial and economic crisis have made the situation worse through the disruption of agricultural supply chains and market, weakening the ability of the agricultural sector to address food security. Despite intervention in agriculture, crop, and livestock production system are still sensitive to drought and other extreme weather events especially in arid and semi-arid areas as indicated by Ndirangu et al. (2017). This confirms the existence of fragile local mechanism for coping and poor resilience to cushion against future climate change shocks. Ochieng (2015) discusses the impact of climate change on agricultural production and considers rainfall and temperature has having positive effects on the revenues of most crops. Fluctuation of temperature and rainfall in growing seasons is known to cause serious problems in normal plant processes, hence crop losses are inevitable. The findings by Ochieng (2015) stressed the significance of the longstanding effects of climate change in terms of temperature changes to have an adverse effect on crop production compared to interim effects. The impacts of temperature in this respect significantly override those of rainfall fluctuations. Fischer (2006) predicted that the global warming phenomenon would lead to higher temperatures and modify precipitation levels which to far extent have had an impact on the productivity of land suitable for agriculture. These are some of the current events being experienced in countries such as Kenya. He further concluded that the integration of agro-ecological zonal methodologies and socioeconomic models could provide right tools for land use planning and resource development.

Farming in most parts of Africa, Kenya included, revolves around agro-ecological zones. It is the *agro-ecological* zones as defined by FAO (1978) and improved by Jiitzold and Kutsch (2000) through the farm management guidelines that determine where crops can be suitably grown. In this agro-ecological zones, cropping seasons have been identified based on the length of growing periods. In addition, FAO (1978) defines agro-ecological zoning (AEZ) as the use of soil parameters, natural features, and climate characteristics to demarcate areas potential for agricultural production. The particular parameters used in the agro-ecological zoning and cropping pattern assign more focus on the climatic and soil, chemical, and physical parameters and their requirements by crops and on the management systems under which the crops are grown. Each zone has a similar definition to ensure its suitability to support certain functions. This kind of evaluation calls for certain recommendations designed to improve the existing land-use suitability situation, either through increasing advocacy for production or by barring use due to land degradation. Moreover, the studies on agro-ecological zoning put emphasis on the use of FAO classification of 1978 which looks at the ability of agricultural land to support crop production through length of growing period, temperature, and precipitation. According to FAO (1978), this length of growing period is understood to be the period where rainfall and stored soil moisture is greater than half of the evapotranspiration. This, therefore, simply means the longer the growing season, the more the plants to be planted, the longer the period for plant carbon fixation, and the higher the yields. Vrieling et al. (2013) indicate that the length of growing period is analyzed based on weather station data, which are unfortunately quite scarce in Africa or depend coarse rainfall satellite data that are not reliable. The findings reported by Vrieling et al. (2013) clearly show that there are high variability in length of growth period in arid and semi-arid areas. This is further confirmed by the high crop failure rate in the same regions.

Length of growth periods derived from earth observation data such as GIMMS Normalized Difference Vegetation Index generation 3 (NDVI3g) dataset provided useful information for mapping of farming system as well as study of climate variability. The other option for assessing the length of growing periods is documented elsewhere (De Beurs and Henebry 2010; White et al. 2009) and consider use of time-series remote sensing data. The remote sensing methods have been found to be convenient in providing useful information for understanding cropping patterns especially at the start and end of seasons. If such information is properly harnessed can provide early warning systems that can be leveraged on for efficient planning. Henricksen (1986) discussed the need for proper definition of agro-ecological zones in Africa. According to him, it would enhance the ease in collecting and updating agro-ecological data and further give a better latitude for proper definition of the cropping seasons.

The role of agro-ecological zone in defining the crop environment in Kenya, has been generated from the agro-climatic zones. An agro-ecological zone is therefore defined by its relevant agro-climatic factors, especially the moisture supply and differentiated as well by soil patterns. The aim is to provide a frame-work for the natural land use potential. The main zones are therefore differentiated by their ability to provide temperature and water requirements for the specific crops in that particular agro-ecological zone. This is related to the climatic yield potentials as calculated from the computer. Once these zones are established, they are comparatively run alongside the Braun's climatic zones of the precipitation/evaporation index as shown by Le Page et al. (2017). The comparability is expected to have variances due to the influence of the length and intensity of arid periods, a factor that has also to be considered. New agro-ecological zoning methodologies have been done and found to be effective in assessing crops against climate vulnerability (Le Page et al. 2017).

Climatic changes have caused drastic weather patterns over the years where more consideration other than length of the growing period are needed. The current length of growing periods are characterized by intensive precipitation whose distribution is poor while at the same time high temperatures are experienced in-between period of precipitation cessation. Inclusion of earth observation data, therefore, becomes very critical in understanding developing climate variability within the existing agro-ecological zones and thereafter necessitating technologies that can accommodate significant changes. Earth observation data and remote sensing technologies have improved and hence provide near real-time observations. Moreover, the placement of rain gauges which determine the precipitation data is sparse and may not reflect microclimatic changes that have occurred over the years as argued by Vrieling et al. (2013). The evaluation of agro-ecological zone focuses on changes in precipitation, temperatures, and soil characteristics which have necessitated the interpretation of changes in growing periods over the years using earth observation data. Furthermore, there has been a pressing need to adapt climate-smart crops for the various regions in Kenya. Coincidentally, climate-smart crop management in the twenty-first century is the new paradigm shift that may be panacea to majority of rain-fed smallholder farmers to cope with climate change. This chapter therefore calls for use of earth observation data such as soil surface moisture data, varied temperature parameters, varied precipitation parameters, and other relevant parameters that can enable the evaluation of cropping seasons based on gridded variation found within the agro-ecological zones from earth observation datasets.

Previous studies have shown that rainfall levels are bound to decrease, increase, or remain the same under most climate scenarios albeit with a lot of extremes in some scenarios, while the temperature are likely to increase. It is these impacts and implications of such reports that have necessitated the need to reevaluate the agro-ecological zonation and infer whether there any significant changes in the growing periods. Ayugi and Tan (2019) revealed a rise in temperature which might contribute to hydrological droughts in the arid and semi-arid areas in future. Additional information have pointed out the need to analyze for specific trends and thereafter characterize areas in terms of vulnerability to climate, particularly the semi-arid areas of Africa, Kenya included. King'uyu et al. (2000) noted significant changes over the surface temperature based on data collected in 71 stations for the period 1939–1992 in Eastern Africa. Unfortunately, geographical variability in nighttime temperature was difficult to interpret thereby calling for further research. Nsubuga and Rautenbach (2018) did a review of climate change and variability in Uganda and confirmed a growing trend in changes in temperature and rainfall and, moreso, in the Eastern Africa region where Kenya and Uganda lie. Hastenrath (2001) and Schreck and Semazzi (2004) have also done extensive work on climate change in East Africa and significantly provided information on changes in weather patterns over years. These reviews of scientific research findings have created a better understanding of the recent climate changes and variabilities in Kenya and provide further information for use in future research and adaptive actions.

The use of earth observation data in climate change monitoring cannot be overemphasized. Earth observation data have provided important insights into important parameters such as biological, physical, and chemical in its bid to address climate change needs and adaptation strategies (Guo et al. 2015). The capacity of earth observation data is advanced in terms of temporal and spatial scales. The complexity in the study of climate change in the aspects of atmosphere, oceans, and lands, make earth observation data an important option as stressed by Guo et al. (2015). The launch of various datasets by NASA and their availability to near real time open a huge platform for addressing climate change issues in Africa. The current chapter attempts to integrate a number of these options to give insights to climate-smart agricultural option in these fragile agro ecological zones IV and V with an aim of providing strong adaptation and adoption strategies to variable weather and cropping seasons.

This chapter focuses on the agro-ecological zones IV and V of the marginal lands of Tharaka-Nithi, Meru, Makueni, and Machakos counties that run along a similar agro-climatic belt in Eastern Kenya. These areas cover semi-arid region where most farmers depend on rain-fed agriculture for livelihood. These areas are characterized by crops as such green grams, sorghum, millet, and pigeon peas. Agro-ecological zones IV and V are classified as a Lower Midlands that are found in semi-arid areas (FAO 1996). According to FAO (1978), the various cropping patterns in the two rainy seasons are evaluated as having the following characteristics:- very short to short and very short cropping seasons, very short to short and very short to short cropping seasons, very short to very short to short cropping seasons, two very short cropping seasons, very uncertain and very short to short cropping season, and finally very uncertain and very short cropping season for Tharaka Nithi. Part of Meru that form part of the agro-climatic belt of Tharaka-Nithi have similar cropping pattern. Machakos' lower midlands zone V is also characterized by very short and very short to short cropping seasons, very short and very short cropping seasons, and finally very short to short and very short to short cropping seasons. Kitui, on the other hand, has very short to short and short to very short cropping seasons, very short and very short cropping seasons, very uncertain and short to short cropping seasons for both zones IV and V (Jiitzold and Kutsch 2000). The average rainfall and temperatures for Tharaka Nithi, Meru, Makueni, Kitui, and Machakos lower midland zones IV and V as documented by the farm management hand book of Kenya is illustrated in Table 1. Illustration of the these zones are also shown in Fig. 1.

Earth Observation Products in Climate Change Monitoring

Time average, soil moisture content underground, land surface skin temperature, multiyear monthly mean surface temperature, air temperature, total precipitation flux, total surface precipitation flux, total surface precipitation, precipitation rate, combined gauge precipitation, monthly precipitation, climatology rainfall flux, and monthly evapotranspiration for varied periods products were obtained from earth observation Geospatial Interactive Online Visualization and Analysis Infrastructure (GIOVANNI) NASA web portal. The products used were precipitation rates in mm/day with a spatial resolution of 0.25° for the period 1998–2019. Precipitation rates here refers to rainfall intensity. Another product used was amount of precipitation on monthly basis recorded as mm/day for the period 1998–2017 packaged as time average maps at 0.5° spatial resolution. Thirdly, soil moisture content underground at 0–10 cm, 10–40 cm, and 40–100 cm was also evaluated. The products were obtained from NOAH through the GLDAS model for the period 2000–2019. Alongside these, the soil temperature at 0–10 cm, 10–40 cm, and 40–100 cm was obtained from Atmospheric Infrared Sounder (AIRS)/National Oceanic and Atmospheric Administration (NOAA) products for the period 1998–2014 at 0.25° spatial resolution. Another temperature parameter examined was the products of air temperature at surface day time and night time as obtained from AIRS/STM for the period 2002–2016 at 1° spatial resolution. Further products on evapotranspiration from NLDAS/MOSO for the period 1998–2014 at 0.25° spatial resolution. The dataset used for this study are general time average maps for periods not less than 10 years to bring various phenomena of variation in the area of study. These phenomena are meant to provide current insight into climate change variability and influence in agro-ecological lower midland zones IV and V. The remote sensing product analyses were done using QGIS, Terrset software, and other GIS software. This was to get current status of the agro-ecological lower midland zones IV and V and whether there are any indications that can lead to change in cropping season and suitability of crops. Other products examined include time average products on amount and distribution of the precipitation as compared to studies done by Jiitzold

Table 1 Characterization of lower midland zone V based on cropping season, annual temperature, and annual average rainfall – Jatzold and Kutsch 2000)

Name of the county	Agro-ecological zones IV and V	Cropping season in the two rainy seasons	Annual mean temperature in °c	Annual average rainfall in mm
Tharaka Nithi and Meru	V	Very short to short and very short to short	24–22.9	800–900
		Very short to short and very short		Too small
		Very short and very short to short		750–870
		Very short and very short		650–850
		Very uncertain and very short to short		630–660
		Very uncertain to very short		600–700
	IV	Short to very short and short to very short	23.5–21.0	820–920
Machakos and Makueni	V	Very short to short and very short to short	24–21.6	650–750
		Very short and very short to short		600–800
		Very uncertain and very short to short		600–700
		Very uncertain to very short		690–700
		Very short and very short		No data

(continued)

Table 1 (continued)

Name of the county	Agro-ecological zones IV and V	Cropping season in the two rainy seasons	Annual mean temperature in °c	Annual average rainfall in mm
	IV	Short to very short and short to very short	22.0–21.0	700–850
		Very short to short and short to very short		700–800
Kitui	V	Very short and very short to short	24.0–23.0	650–790
		Very short and very short		600–780
		Very uncertain and very short to short		600–750
		(Very short) and very short		600–650
		Very uncertain and very short		550–630
	IV	Short to very short and short	24.0–20.9	800–1000
		Very short to short and short to very short		750–880
		Very short and short to very short		700–820
		Very uncertain and short to very short		720–820

and Kutsch (2000) that are mostly used for national policy planning on cropping season in Kenya through the farm management guideline books.

Climate Variability in Agro-ecological Lower Midland Zones IV and V Soil, Air, and Land Surface Temperature Variations

This chapter gives a brief overview of air temperatures, land surface temperature, and multiyear land surface skin temperature daytime for the period 2000 to 2016 that depict an upward trend in temperature changes (Figs. 2 and 3, respectively) for zones IV and V. These areas that include parts of Makueni and Kitui show temperatures of between 26 °C and 31°C while Machakos, Tharaka, and parts of Meru have temperature ranges of 24–26°C. These temperatures changes compared closely with those earlier reported by Jiitzold and Kutsch (2000) (Table 1) with annual average temperature ranges of 22–24 °C for Machakos, Kitui, and Makueni indicating an increasing trend in temperature. Tharaka Nithi and parts of Meru temperature fluctuate between 24–26 °C and 22–24 °C, respectively. The trend of temperature

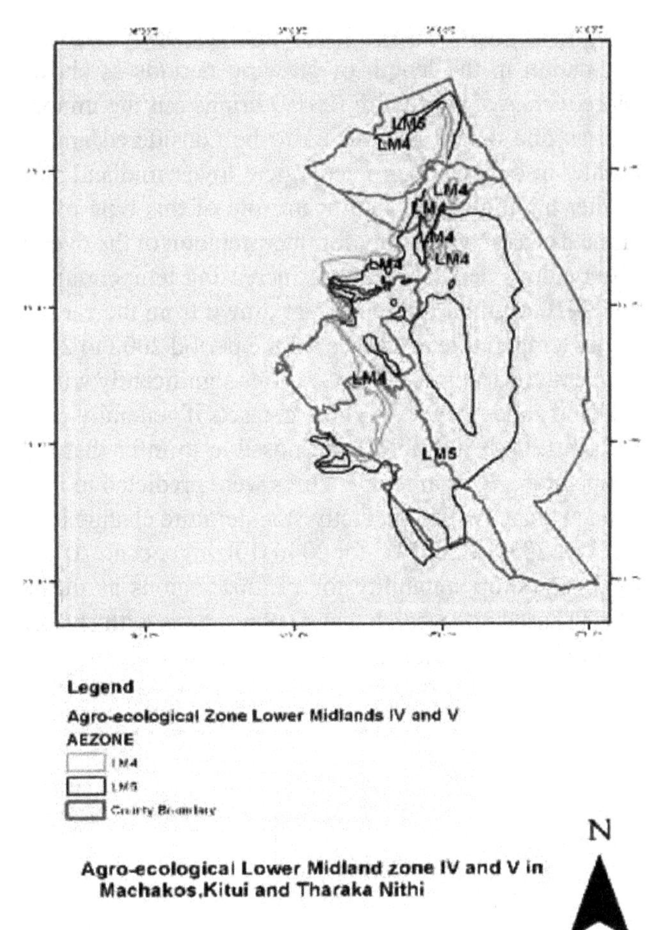

Fig. 1 Agro-ecological lower midland zone IV and V in Machakos, Kitui and Tharaka Nithi

variations confirms reports by previous studies (Nsubuga et al. 2014; Omondi et al. 2014), an indication of increase in temperature trends and variability over East Africa. The challenge of variations from one locality to another as highlighted by Omondi et al. (2014) may be solved through having gridded earth observations data. Earth observation data provide spatial resolution smaller compared to data collected from various in situ stations whose spatial resolutions are usually very wide due to unavailability of adequate and well-distributed weather or radar station in Kenya. There are pockets of temperature variations as shown on the time-average maps of air temperature for the period 2002–2016 which is a 14-years-period for both daytime and nighttime (Fig. 2). This is an indication that over the years, different areas within the agro-ecological zones of lower midland IV and V have experienced climate variability in terms of temperature changes. This may also imply that the length of growing periods of crops are different within the same zone. Vrieling et al. (2013) is of the opinion that the distribution of crop and farming systems go hand in hand with the length of growing period. They further note that length of growing period in Africa (Kenya included), have been determined by weather station data with poor spatial distribution.

Climate change over the years has been studied and confirmed to cause a fluctuation in the length of growing periods as shown by Gregory et al. (2005). Moreover, Ayugi and Tan (2019) brings out the importance of surface air temperature as one of the parameters to be considered and blends well with the current finding in regard to agro-ecological lower midland zone IV. Ayugi and Tan (2019) further highlights the significant role of this type of products in the assessment of climate change variability for interpretation of the overall climate state in Kenya. His observations depict a trend of increasing temperatures in the period between 1971 and 2010. Similar inferences are drawn from the earth observation climate products for air temperature at surface for the period 2002 to 2016 in the similar areas. The air temperature at surface (Fig. 2) varies significantly within the agro-ecologies of lower midland zones IV and V. These datasets if generally compared to data by Jiitzold and Kutsch (2000) in Table 1 are possible to infer that there is an increasing trend in temperature. Rise in temperatures were predicted to increase between 1 °C and 2.5 °C on average, with a predictive temperature change in 2020 being between 1 °C and 2 °C for 2030 and 2.5 °C for 2040 (Eitzinger et al. 2011). These predictions are likely to change crop suitability for various regions as further argued by Eitzinger et al. (2011) based on research that displays maps with changing suitability of tea growing area in Kenya. He further cautions that areas between 1400 and 2000 meters above

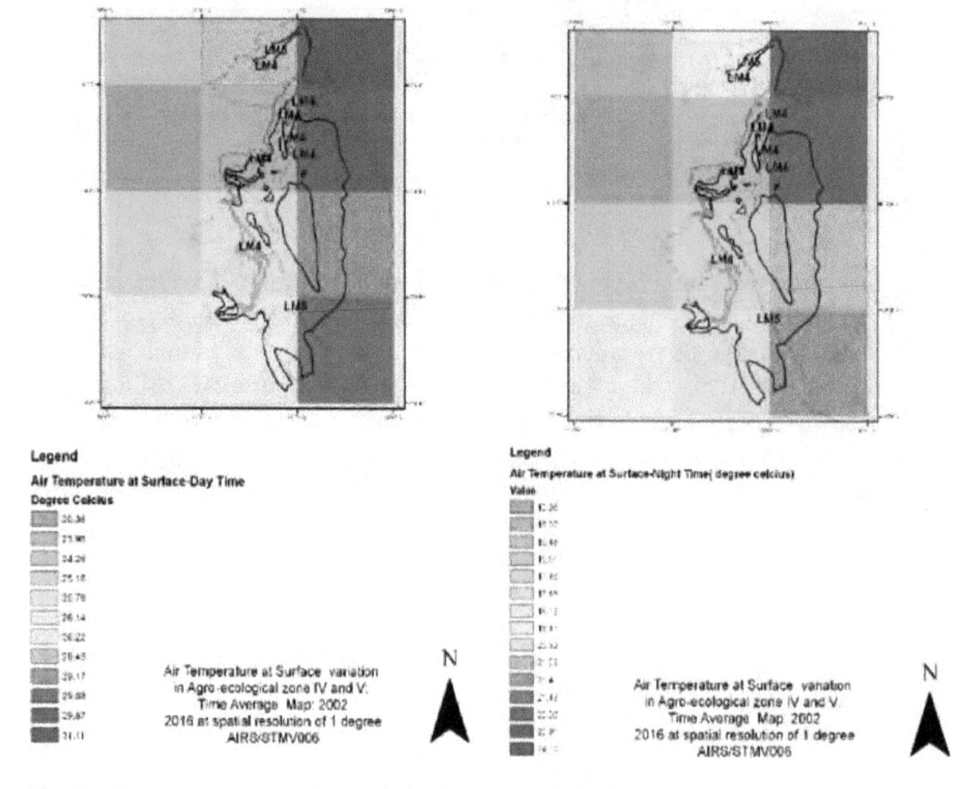

Fig. 2 Air temperature at surface variation in agro-ecological lower midland zone IV and V

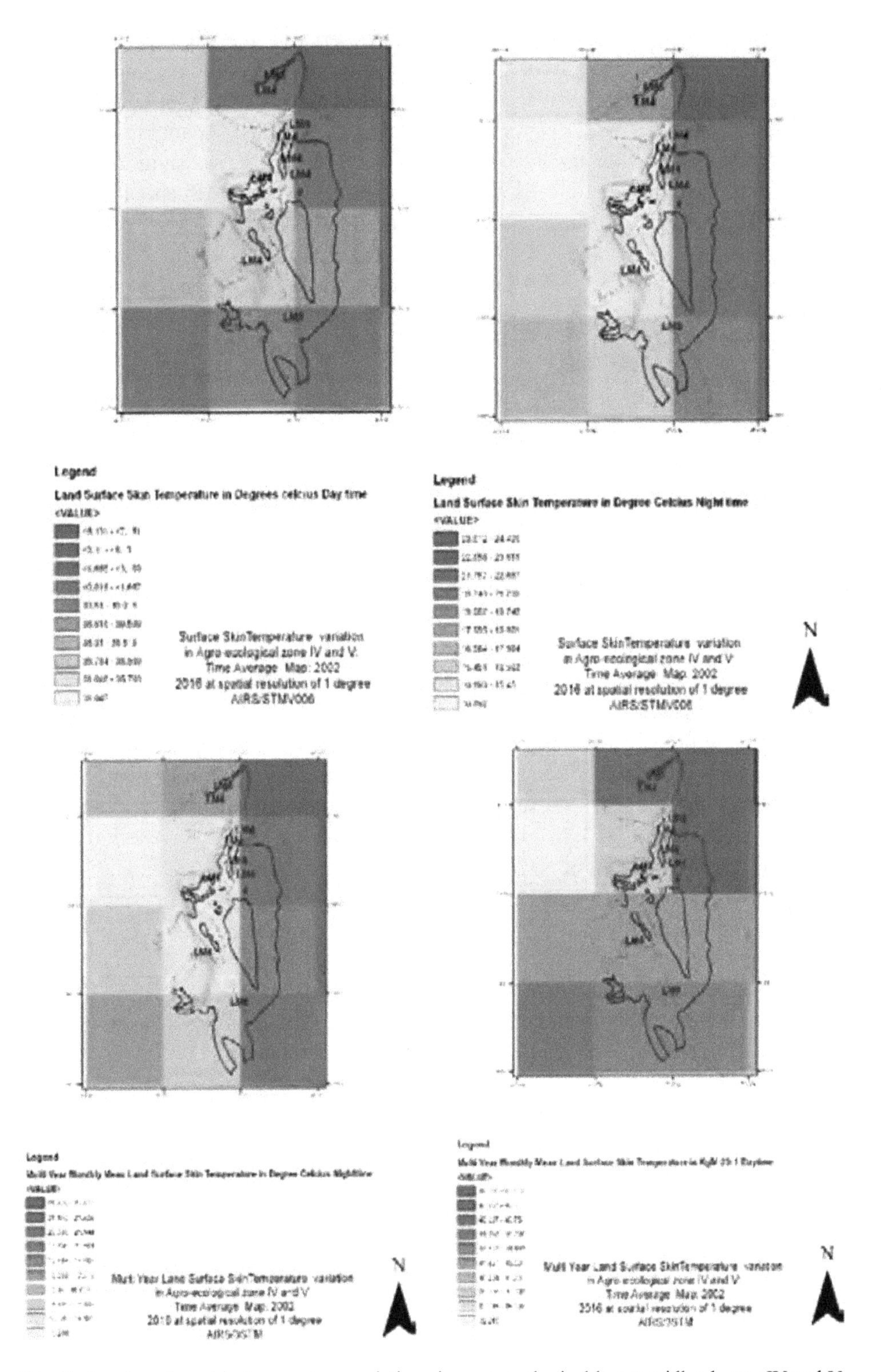

Fig. 3 Land surface skin temperature variations in agro-ecological lower midland zone IV and V

sea level are likely to be regions that will experience the highest changes in suitability for crops.

Additionally, this chapter made effort to deeply make observations on monthly Land surface skin temperatures for the period 2002–2016 at 1° spatial resolution to further improve on understanding of the analysis. Figure 3 shows significant variations within the agro-ecological lower midland zones IV and V. Such variability in land surface temperatures in any area is a manifestation of variation of vegetation responses to water stress (Sun 2009; Pinheiro et al. 2006). Land surface temperature has been confirmed to be a very important parameter for environmental as well as climate studies (Pinheiro et al. 2006). It is also important to note that currently, weather stations provide limited data on spatial patterns of temperatures over large areas as shown by Assiri (2017). This leaves interpretation of the land surface temperatures from satellite imagery a viable option. It would, therefore, be prudent to argue that land surface temperature is a key parameter in understanding crop water stress within any agro-ecological zone.

The significance of land surface temperatures in climate monitoring cannot be ruled out in the extreme weather event scenarios as reported by Sun (2009). Furthermore, Vlassova et al. (2014) emphasize land surface temperature as an important parameter in soil-vegetation transfer modeling in most terrestrial environments. This parameter becomes even more important in evaluating cropping seasons and assessment of length of growing periods in such areas. Time average maps of multi-year monthly daytime and nighttime land surface temperatures for the period 2002–2016 are also presented for agro-ecological zones IV and V (Fig. 3). Evidently, there are significant variations across the entire agro-ecological lower midlands zones IV and V area as also confirmed by the monthly daytime and nighttime temperatures for the same period. Assiri (2017) demonstrated high correlation between moderate resolution imaging spectroradiometer (MODIS) satellite nighttime land surface temperature which also had high correlation with station-based minimum temperatures as compared to the daytime land surface temperatures. The accuracy and effectiveness of MODIS land surface temperature was further corroborated by those of Kenawy et al. (2019). Apparently, Sun (2009) showed that land surface temperature uses/values to be quite wide and included but not limited to hazard prediction, water management in agriculture, crop management, in terms of crop stress monitoring and yield forecasting and nonrenewable resource management. In addition, the authors provide contribution to work done regarding assimilation of land surface temperatures in soil moisture monitoring through the surface energy balance assimilation scheme.

This chapter also makes observations on soil temperatures for 0–10 cm, 10–40 cm, and 40–100 cm for the period 1998–2014 at a spatial resolution of 0.25° since the focus of understanding agro-ecological zones currently suitable for certain crops and the cropping seasons. These products do not seem to show significant differences among the various soil depths (Fig. 4). This could probably be attributed to vegetation cover, an important element in thermal conditions of soils (Holmes et al. 2008). This condition has the ability to influence the vertical differentiation of soil temperatures as previously revealed by Skawina et al. (1999). What was striking is

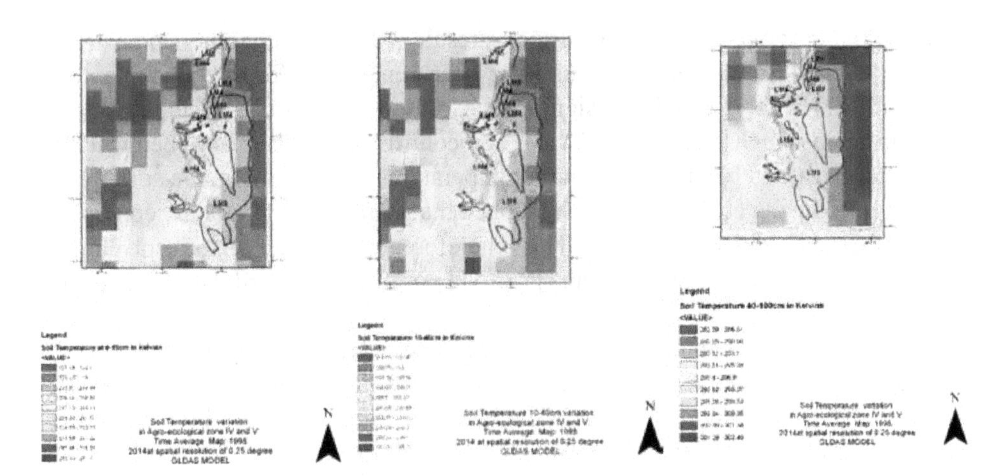

Fig. 4 Soil temperature variations in agro-ecological lower midland zone IV and V

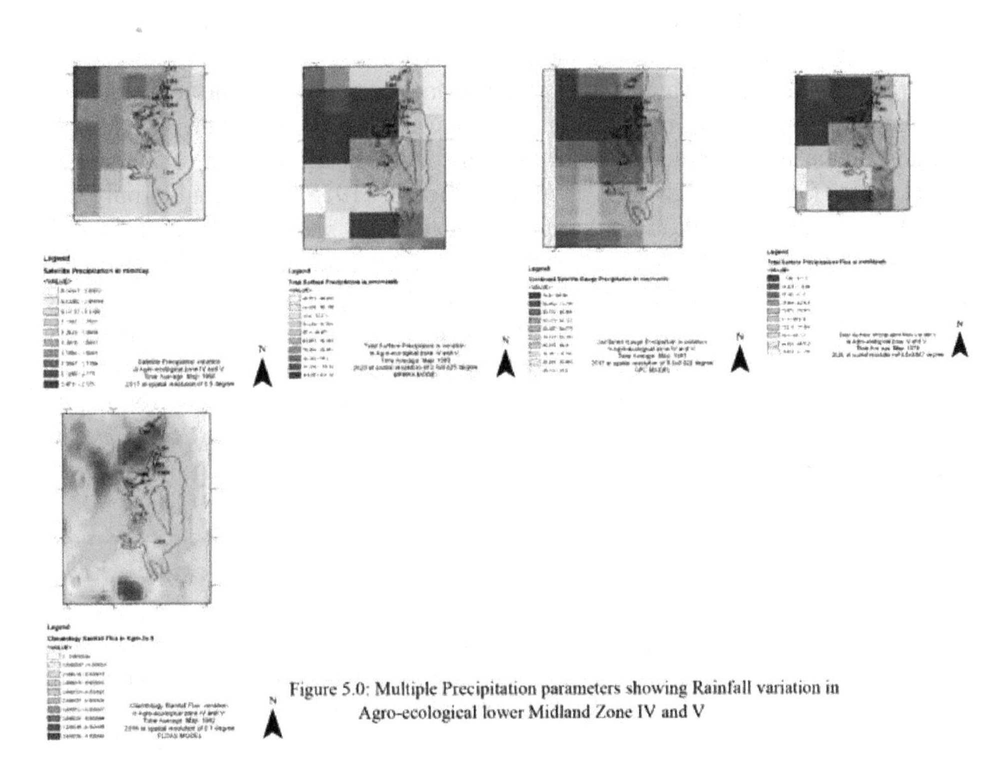

Fig. 5 Multiple precipitation parameters showing rainfall variation in agro-ecological lower midland zone IV and V

the significant variation identified within the agro-ecological lower midland zones IV and V with the soil temperatures varying across Makueni, Machakos, Tharaka Nithi all the way to the small parts of Meru, i.e., the ranges are 26.85–22.85, 20.85–24.85, 24.85–26.85, and 17.85–23.85, respectively, for all areas (see Fig. 5). These observations summarize quite significant and clear changes that have been taking place within the zones over a long period as result of climate variability and extreme weather events. Similar observations have been reported before by Buckman and Brady (1971) who stressed the significance of soil in agriculture since it is the medium upon which crops are grown. They confirm that significant chemical and biological reactions begin when certain temperatures are optimal in the soil media. Moreover, Skawina et al. (1999) demonstrated the significance of weather conditions on the thermal properties of soil. This, therefore, means that significant climate variability influence soil temperatures positively or negatively and thereby affecting crop growth.

Soil temperature parameters therefore become extremely important factors to consider in agro-ecological zone characterization as shown by studies of Osińska-Skotak (2007) who reported that soil temperatures were influenced by meteorological situations taking place. Onwuka (2018) confirms that soil temperatures are affected by environmental factors especially by controlling the heat on the soil surface and heat dissipated from the soil down the profile. The importance of soil temperature cannot be overemphasized in this chapter. Indeed Sabri et al. (2018) showed that biological processes such as seed germination, seedling emergence, plant root growth, and the availability of key nutrients are components dependent on soil temperature, moreso the transmission of water inside the plant. In addition, injuries in the plant tissue at root zone can occur as a result of fluctuation in temperatures (lower or higher) as attested by Decker (1955). This could imply that soil temperatures fluctuation could the reason for high soil-borne diseases due to injury of plant tissues. It could also be the explanation for high soil-borne disease incidences associated with climate change. Schollaert Uz et al. (2019) documents the role that remote sensing datasets can play in linking different climatic variability events to pest and disease incidences. In this chapter, the coupling of crop models to pest and diseases model together with climates are considered as new frontiers to enabling monitoring of pests and diseases outbreaks under extreme weather events for early preventative measures.

Precipitation Amount, Rate, and Soil Moisture Variations

Several earth observation products modeled for precipitation analysis were used in this chapter for understanding variations within the agro-ecological lower midland zones IV and V as shown in Fig. 5. The first product we considered was the average time maps for precipitation monthly for the period 1998–2017 (recorded in mm/day) at spatial resolution of 0.5°. This revealed variations in precipitation amounts within these areas. The amounts varied across the areas in consideration with Kitui and Makueni showing a range between 1.39 and 0.9 mm/day, while Machakaos had precipitation amounting to 1.59–1.89 mm/day and Tharaka Nithi and Meru exhibiting 1.39–1.89 mm/day and 1.26–1.59 mm/day, respectively. Another product of interest that observed was the total monthly surface precipitation in mm/day for a spatial resolution 0.5 × 0.625° for the period 1980–2020. This product also showed significant variation within the agro-ecological lower midland zones IV and V where places like Makueni and Kitui showed precipitation amounts of 63.67–44.43 mm/month with small pockets showing higher precipitation amounts of 70.36–82.49 mm/month. Machakos area had rainfall ranges between 70.36 and 94.2 mm/month with some small areas to the North East experiencing precipitation amounts of 94.2–118.04 mm/month. Finally, Tharaka Nithi and parts of the drier Meru experienced precipitation amounts of 70.36–118.04 and 53.4–48.2 mm/month, respectively. The combined satellite gauge precipitation monthly in mm/month at spatial resolution of 0.5° for the period 1983–2017 also formed an interesting observation set. There were clear and interesting variations on precipitation amounts within the agro-ecological lower midland zones IV and V. This product specifically showed 49.17–62.72 mm/month for Makueni and Kitui, while for Machakos the precipitation ranges between 69.96 and 101.61 mm/month. Tharaka Nithi and Meru show precipitation amount of 59.77–101.61 mm/month with a small area to the northeast showing 41.51–49.17 mm/month. Finally, the time average maps of climatology of rainfall flux monthly for the period 1982–2016 at spatial resolution of 0.1° in $Kg/m^{-2}/s^{-1}$ was also observed. The observation pointed to the fact that substantial variations existed within the area of the agro-ecological lower midland zones IV and V. All products showed a significant rise in precipitation amount for areas, especially Makueni, Kitui, and Machakos. We observe that these areas have experienced gradual increase in rainfall amounts in some pockets, e.g., Machakos as compared to earlier reports by Jiitzold and Kutsch (2000), implying a rising trend. On the contrary, Makueni and Kitui areas seem to register a downward trend in precipitation (see Fig. 5 and Table 1). Over the years, these semi-arid regions have been described to be regions of limited precipitation (Slatyer and Mabbutt 1964). However, Managua (2011) had conflicting views, indicating a rise in rainfall for the period 2020–2050 for most parts of Kenya. These scenarios of high unprecedented rainfall are currently being experienced in Kenya.

Precipitations in the marginal areas have been described to undergone considerable positive change, indicating trends of increasing rainfall amounts in some pockets. Other studies have reported conflicting (negative trends), indicating the failure to include the changing Indian Ocean weather patterns; – something that may affect climate data modeling along semi-arid regions as argued by Herrero et al. (2010). Vrieling et al. (2013) gave a comprehensive descriptive dataset on changes in weather patterns which showed an increasing trend in the dry period between the short and long rains and this has been happening at the expense of short rains. Interestingly, Elbasit et al. (2014) has also demonstrated that there is a good agreement between satellite Tropical Rainfall Measuring Mission (TRMM) 3B43 products and the monthly rain gauge information, confirming increasing trend in rainfall and the reliability on Tropical Rainfall Measuring Mission (TRMM) earth observation products for climate monitoring.

The variations observed and scenarios of higher precipitations in some part of the agro-ecological lower midlands zones IV and V call for future evaluation of the cropping season as well as the length of growth periods. The proper and reliable estimation of onset and cessation of precipitation is critical in rain-fed agriculture in the semi-arid areas as emphasized by Fiwa (2014). Incidentally, existing information shows that the relationship between precipitation onset, cessation, and length of growing period becomes very important in the planning of agricultural activities especially among smallholder farmers in Africa (Fiwa et al. 2014). In this chapter, therefore, we are of the opinion that there is an upward increase in precipitation in these marginal areas and but the absolute increase the precipitations have vast variability. Aming et al. (2014) also identified precipitation extremes and higher variabilities in arid and semi-arid parts of Africa, the findings which seem to be in line with our observations in these agro-ecological lower midland zones IV and V of Kenya. Fluctuation and variations in the rainfall have also been confirmed by Camberlin (2009) with indications showing serious variations from one weather station to another, especially in the long rains of East Africa. The growing developments in satellite-based rainfall assessments provide a cheaper alternative to rainfall data that is available for free online (Kumar and Reshmidevi 2013). The collection of rainfall data using passive or active remote sensing techniques has a potential to bring a more informed way of handling climate change in Africa and moreso Kenya with huge landmass represented by arid and semi-arid regions. These methods of rainfall data collection can be described as brightness temperature for passive method (Hengl et al. 2010) and attenuation of the radar power at several heights to estimate surface rain for the active method (Skolnik 1962; Meneghini et al. 1983; Hengl et al. 2010).

Rainfall intensity from earth observation data is known to show variations within the agro-ecological lower midland zones IV and V (Fig. 6). Higher precipitation rates are associated with areas of high precipitation as shown in Fig. 6. High precipitation rate is associated with runoff and hence low soil moisture content (Mutiga et al. 2013). This further affects the length of growing period and expected cropping seasons. Notably, high runoff is associated with low productivity of soil and causes severe risk in agriculture as result of climate hazard (Gatot et al. 2001). Furthermore, it is known that soils experiencing high runoff as result of high rainfall intensity are normally very low infertility due to the percolation of soil nutrients. Rainfall intensity monitoring in agro-ecological zone of marginal areas of Kenya is key since high level of runoff in the semi-arid areas are highly associated with precipitation rates as well as poor farming practices. Fiwa et al. (2014) asserts that rainfall intensity monitoring using weather satiation is a real challenge across Africa. Thies (2008) evaluated rainfall intensity data from satellite imagery differentiation techniques and concludes that they offer potential for improved rainfall rate. This allows for spatiotemporal near real-time information on rainfall distribution. This kind of data and techniques in Africa will definitely address the challenges facing climate-smart crop management under extreme weather events whenever they will be fully available.

Additional products considered included soil moisture content below the ground at 0–10 cm, 10–40 cm, and 40–100 cm, soil temperatures at 0–10 cm, 10–40 cm, and 40–100 cm, see (Fig. 7). These additional products provide information on moisture availability for the various crops within the agro-ecological lower midland zones IV

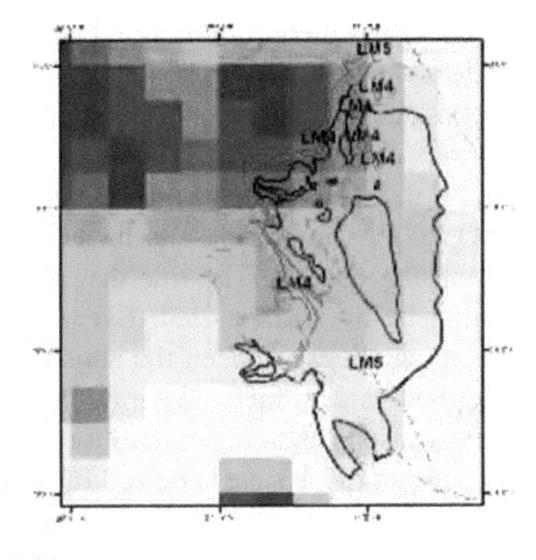

Fig. 6 Daily precipitation rate variation in agro-ecological lower midland zone IV and V

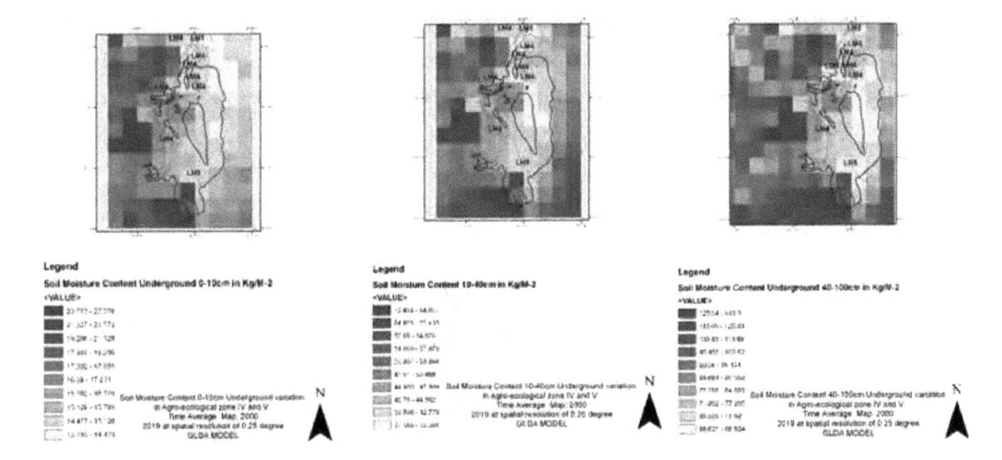

Fig. 7 Soil moisture content underground variations in agro-ecological lower midland zones IV and V

and V. This chapter makes a contribution on knowledge on increased variability in soil moisture content across the entire study area. This can be attributed to the verse variance in soil types and soil characterizes as well as canopy cover. This can also be related to the various changes in climatic;- soil related weather parameters. Other factors affecting soil moisture contents are human-induced activities that cause land degradation (FAO 2002). Degraded soils have very low water retention capacity hence low soil moisture content. The role that soil moisture content plays in crop development cannot be overemphasized. Every crop requires good soil moisture content to stimulate activities between the shoot and the root zone. Soil moisture contents are identified as key factor determining length of growth period which impact on cropping seasons and crop suitability in any agro-ecological zones.

Kamara and Jackson (1997) states that soil moisture is a better indicator of water availability to the crop than rainfall amount. Improved soil moisture management can only take place when analysis of soil moisture content are carried out and are well understood. Understanding soil moisture content scenarios assist in working toward high water retentions in the soil which eventual reduction in soil runoff in some cases. In semi-arid regions, the issue of soil moisture becomes very important since moisture stress in crop is one of the key determinants of crop failure. In other studies, Pellarin et al. (2020) has shown that soil moisture content can be used to infer precipitation using PrISM Model. This kind of modeling is said to provide useful information concerning crop yield estimates and irrigation demands over large areas. In his research, comparison made with weather station information showed that there was high correlation proving the significance and the role that soil moisture content below ground plays in understanding shifting precipitation patterns. Overall soil moisture content is vital in understanding cropping season, i.e., low soil moisture levels are indications of declining length of growing period for crops. The biggest challenge identified by this chapter is existence of many soil moisture algorithms and products that cannot be compared to other parameters such as precipitation as propounded by Pellarin et al. (2020). He further indicated the need for analysis that can integrate other parameters such as precipitation for the proper validation of satellite soil moisture and also system that can determine the relationship between different soil moisture products.

Evapotranspiration Variations

The other parameter that this chapter observed is evapotranspiration. The monthly evapotranspiration earth observation reported here are for the period 1998–2014 at spatial resolution of 1° and recorded in $Kgm^{-2}\ s^{-1}$ (see Fig. 8). The values indicated a variation in evapotranspiration across the agro-ecological lower midland zones IV and V area. There were no products for comparison with previous periods in order to understand any changes in evapotranspiration. Areas like Meru agro-ecological lower midland zones IV and V revealed higher evapotranspiration rates. This could be explained by high vegetation cover known to be in this region and high soil moisture content. Evapotranspiration is directly related to soil moisture content and

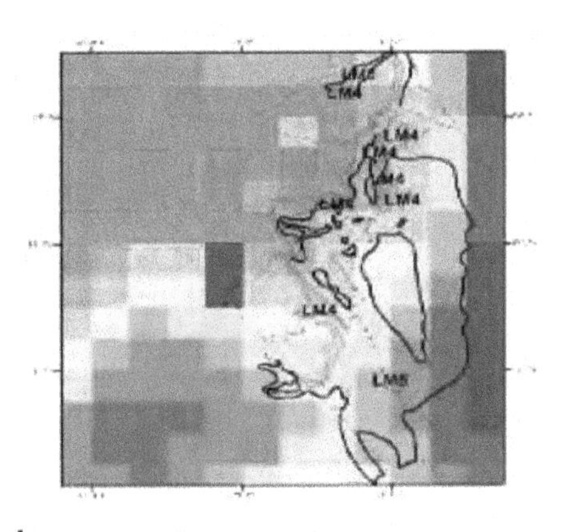

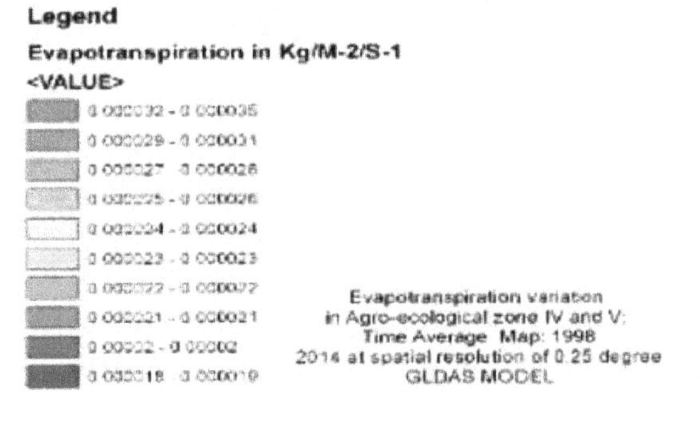

Fig. 8 Evapotranspiration variation in agro-ecological lower midland zone IV and V

atmospheric moisture demand (Fischer et al. 2008). Further Fisher et al. (2008) shows that soil moisture is as a result of precipitation while atmospheric moisture emanates from radiation that is controlled by surface and atmospheric temperatures.

Areas in Kitui and Makueni experience very low Evapotranspiration, which could also be due to low vegetation covers that are characteristic of these areas. These indicate that areas of high evapotranspiration rates are likely to have longer growing period. Table 1 classifies these areas and reveal that they have similar short cropping season as those with very low evapotranspiration rates. Evapotranspiration as shown by studies of Jiitzold and Kutsch (2000) shows variations in the length of growing periods which eventually determine the cropping seasons. In this case, the evapotranspiration employed statistical methods of Penman & McCulloch with albedo for green grass being 0.2; McCulloch (1965). This is not to imply its inadequacy but rather the need for improvement. Liou and Kar (2014) confirms that evapotranspiration at a global or regional scale can be done by combining surface parameters obtained from remote sensing data and surface meteorological variable and vegetation characteristics. Marshall and Funk (2014) describes evapotranspiration as an important component in the energy, water, and geochemical cycles that influence climate properties and further says that its interaction with drivers of climate change remain unexplored in Africa.

A combination of a number of parameters and variables in a climate change scenario is likely to improve evapotranspiration analysis and interpretation. This would further improve on the assessments of cropping seasons. Marshall and Funk (2014) in his discussion points out the usefulness of satellite imagery in representing important characteristics of evapotranspiration over Africa. Similar results have been reported by Dai (2010) who showed rising temperatures in Africa and a decreasing evapotranspiration as result of low precipitation and low soil moisture content. Satellite imagery products through available remote sensing techniques can enhance the availability of spatial and temporal datasets of Normalized Difference Vegetation Index (NDVI), Leaf Area Index (LAI), fraction of incidence sunlight that reflects, surface radiations, and radiometric surface temperatures that are indirectly related to evapotranspiration (Liou and Kar 2014). Further Henricksen (1986) present extensive report on the length of growing period model that was used that seem to work well in drought simulations in Africa. He pointed out the inability of the model to analyze moisture level below 1 month and the assumption of the model concerning runoff and deep percolation.

Contribution of GIS and Remote Sensing in Agro-ecological Zone Evaluations

The suitability of gridded climate indices in monitoring climatic variation is demonstrated in Donat et al. (2013) studies. The gridded climatic indices can enable near real-time monitoring of events as well as their placement for long-term use at both global and regional scales. Boitt et al. (2014) in his studies on the impact of climate change on agro-ecological zones and this chapter has made quite a number of inferences from their work regarding GIS and remote sensing. There are siginificant shifts in agro-ecological zones based on datasets analyzed and projections for future in 2050. These shifts and changes can be picked at apporoximately 1 km and significantly identified as "zone shift." Boitt et al. (2014) further concluded that multivariate clustering under a GIS enviroment is a very informative tool for agro-ecological zone definition. Global agro-ecological zone module V is the latest model for defining crop suitability for various regions. This module utilizes various climate datasets such as number of rainy days; mean minimum, mean maximum temperature; diurnal temperature range; cloudiness; wind speed and vapor pressure (IIASA/FAO 2012). Among this climate dataset, soil parameters are accommodated extensively (IIASA/FAO 2012). Its use among African countries is still limited and therefore needs to be scaled for adoption to the benefit of smallholder farmer levels. Its adoption is generally challenged by the limited use of GIS and remote sensing datasets in most African countries (Rowland et al. 2007). This module could provide the new insights into understanding climate variability and its impact on cropping seasons. Its use in evaluating existing information on agro-ecological zones of lower midland IV and V is paramount.

Wango et al. (2018) explore the suitability of WorldClim dataset in climate analysis and monitoring in the advent of climate change. All previous studies highlighted the role of GIS and remote in transforming the understanding of climate variability, thereby successfully addressing the climate change issues in Kenya and

Africa at large. For example, Henricksen (1986) emphasize the use of satellite remote sensing to capture and monitor the climate variability occurring year after year. Another research by Bartoszek et al. (2015), for Poland, affirms the use of satellite data as useful source of temporal and spatial variability of information on climate. FAO (2017) in the report on review of remote sensing tools, products, and methodologies points out, the issues that Africa is facing, when it comes to improved crop production forecasting. The reports highlights the improvement in technologies related to remote sensing and the mode of communication which are at varying levels of development. Further, the reports mention the problems of various institutions, especially the government, to integrate and use effectively remote sensing products. This in the end has affected timely crop forecasting across the continent. Finally, there are readily available good products of early warning system that Africa should take advantage of since they are free.

Climate-Smart Crop Management

According to Mungai (2017), smallholder farmers in the East African region are facing unprecedented challenges in pursuit of increased production under increased climate change and variability. His work further gives highlights on the dire need for information on possible risk and viable management strategies. Information on risk stem from understanding the climate variability within the agro-ecological zones where farming is defined as discussed in previous sections of this chapter. Nsubuga and Rautenbach (2018) concluded that climate variability is bound to have considerable effects in terms of food availability, especially on the agriculture depended populations. They further point out that climatic variations and differences will continue to have a significant role in the geographic distribution of crop production. Understanding climate variations within agro-ecological zones and their influence on the soil environment is key in climate-smart crop management. The parameters selected in this chapter present time average maps that have shown increasing variations over years as result of climate change. Scaling down satellite imagery and remote sensing technologies for use by smallholder farmer presents one possible solution for the reduction of the impact of climate variability on food security. Previous studies have presented the challenges therein while highlighting various solutions that can be adopted. Nsubuga and Rautenbach (2018) in their reviews highlighted the importance of rainfall measurements in Uganda but failed to show a downward trend in rainfall amount yet there has been as shift precipitation rates as well as end and start of seasons. We looked at some of the climate-smart crop management practices that have been suggested and implemented. The chapter underpins the importance of climatic variability evaluation at agro-ecological zone level scaled down to farmers needs for successful climate-smart crop management. Food and Agriculture Organization (FAO) focus on climate-smart crop production pushes for crop production and practices that enable climate change and mitigation. In this view, different methods have been recommended and used. One of the approaches that has been advocated for is cropping systems which are believed to cushion farmers against climate change shocks. The other is the use of quality seeds and plantings adapted to the various environments. Sustainability and the resilience of a production system is believed to be achieved through improved crop varieties suited for a wide range of agro-ecosystem FAO (2013).

Changes in temperature, precipitation, assessment of evapotranspiration characteristics and soil temperature-moisture regimes necessitate for new frontiers in the management of crops through introduction of adoptable technologies. Report by FAO (2013) on climate-smart crop production pushes for agricultural systems that are very efficient in terms of inputs while at the same having less variability. This will enable sustainability through the stability in the outputs hence more resilient in nature and therefore able to cushion against climate change shock and long-term variability. On the other hand, climate-smart intervention by smallholder farmers cannot be ignored (Ullah et al. 2019). Even though most of the local coping strategies and mechanism may be weak as earlier mentioned, they can be strengthened by incorporating them in technology-based smart inventions.

Another adaptation for climate-smart agriculture is through crop modeling that focuses on building early warning monitoring systems that effectively alter the overall management of the crop in volatile climate environment (Ullah et al. 2019). In his studies, he highlights the use of weather data smart interventions that involve close monitoring of climate variability and relaying of information to relevant stakeholders. This is bound to improve further if implemented under an environment of well evaluated and monitored agro-ecological zoning. This means, the cropping seasons and length of growing period variations within agro-ecological zones will be factored in. The advent of weather data smart intervention and yield forecasting through crop modeling place the use of satellite imagery in products provision in crucial place. Weather data smart intervention cannot forget platforms of forecasting more frequently to the farmers. Stigter (2010) gave focus on the importance of such advisory services in his studies. Zuma-Netshiukhwi et al. (2016) in his study emphasizes on the need for agro-meteorological knowledge transfer or extension services to end-users such as farmers and other relevant stakeholders. Further, he discusses the need for downscaling of seasonal climate to lower resolutions that can address the farmers' needs. Abura (2017) argues that there is need for meteorological department to streamline climate advisory services to the locals so that livelihoods risk as result of climate change can be reduced. This can be done by training of agricultural extension officer who in most developing countries work hand in hand with the smallholder farmers.

Conclusion

This chapter concludes that variation within agro-ecological lower midland zones IV and V does exist and changes have been occurring over a long period of time. All the products from GIOVANNI NASA Earth data website have shown enormous variations. Secondly, earth observation products used alongside in situ information have the potential of improving agro-ecological zoning and the interpretation of the cropping season and crop suitability. Thirdly, climate studies that have relayed conclusive information on agro-ecological zones have focused more on rainfall with exclusion of more datasets that can provide more insight in terms of understanding climatic trends and cropping systems.

Fourthly, use of more varied datasets that focus on varied precipitation parameters, varied temperatures parameters, evapotranspiration, thermal conditions of soil, and soil moistures content are likely to improve on the understanding of length of growing period, hence enabling the adoptions of crop varieties that are climate-smart for particular regions. The low adoption of earth observation products and technologies in remote sensing poses a problem in the monitoring of climate variability in Africa. Effective assessment of length of growing periods from climate information for Africa need to be strengthened and improved through adoption of earth observation products alongside in situ surveillance. Finally, the climate-smart crop management must be considered to cut across the various sectors and not agriculture alone if successful implementation is to be achieved. Agricultural systems can achieve climate-smart objectives through continuous monitoring of activities being undertaken at the agro-ecological zones level since they form very important platforms of crop productions in Kenya. This study confirms that climate-smart crop management under extreme weather in Kenya is unavoidable. This management will stem from the adoptions of technologies, strategies, and policies that focus on climate monitoring, evaluation of exiting agro-ecological zones, and adoption of agro-ecosystems that can support sustainable agriculture. In addition, future studies on climate change impact on marginal areas should consider focusing on further reclassification of agro-ecological zones in Kenya and evaluation of cropping season using both satellite imagery and in situ information to compensate for challenges and gaps experienced in previous studies. This will assist smallholders' farmers in addressing the current challenges that they have as a result of climate variability. Soil moisture assessment and integration and its inter-comparison to other parameters still remain a challenge that needs to be addressed.

References

Abura BA, Hayombe PO, Tonui WK (2017) Rainfall and temperature variations overtime (1986–2015) in Siaya county, Kenya. Int J Educ Res 5:11–20

Aming P, Awange JL, Forootan E, Ogallo A, Girmaw B, Fesseha I, Kululetera V, Mbati M, Kilavi M, King M, Adek P, Njogu A, Badr M, Musa A, Muchiri P (2014) Changes in temperature and precipitation extremes over the Greater Horn of Africa region from 1961 to 2010. 1277(June 2013): 1262–1277. https://doi.org/10.1002/joc.3763

Assiri ME (2017) Assessing MODIS land surface temperature (LST) over Jeddah. Polish J Environ Stud 26:1461–1470. https://doi.org/10.15244/pjoes/68960

Ayugi BO, Tan G (2019) Recent trends of surface air temperatures over Kenya from 1971 to 2010. Meteorol Atmos Phys 131:1401–1413. https://doi.org/10.1007/s00703-018-0644-z

Bartoszek K, Siłuch M, Bednarczyk P (2015) Characteristics of the onset of the growing season in Poland based on the application of remotely sensed data in the context of weather conditions and land cover types. Eur J Remote Sens 48:327–344

De Beurs KM, Henebry GM (2010) Spatio-temporal statistical methods for modeling land surface phenology. In: Hudson IL, Keatley MR (eds) Phenological research: methods for environmental and climate change analysis. Springer, Dordrecht, pp 177–208

Boitt MK, Mundia CN, Pellikka P (2014) Modelling the impacts of climate change on agro-ecological zones-a case study of Taita Hills, Kenya. Univers J Geosci 2:172–179. https://doi.org/10.13189/ujg.2014.020602

Buckman HC, Brady NC (1971) Soil and its properties (in polish). PWRiL Press, Warsaw

Camberlin P, Moron V, Okoola R et al (2009) Components of rainy seasons' variability in equatorial East Africa: onset, cessation, rainfall frequency and intensity. Theor Appl Climatol 98:237–249. https://doi.org/10.1007/s00704-009-0113-1

Dai A (2010) Drought under global warming. Interdiscip Rev Clim Change 3(6):617–617. https://doi.org/10.1002/wcc.81

Decker WL (1955) Determination of soil temperatures from meteorological data. Retrospect Teses Dissert 13252. https://lib.dr.iastate.edu/rtd/13252

Donat MG, Alexander LV, Yang H et al (2013) Global land-based datasets for monitoring climatic extremes. Bull Am Meteorol Soc 94:997–1006. https://doi.org/10.1175/BAMS-D-12-00109.1

Eitzinger A, Laderach P, Quiroga A, et al (2011) Future climate scenarios for Kenya' s tea growing areas. pp 1–27

Elbasit MA, Adam EO, Khalid Abu-Talib1, Ahmed F, Yasuda H, Ojha CSP (2014) Validation of satellite-based rainfall measurements in arid and semi-arid regions of sudan. Proc 10th Int Conf AARSE

FAO (1978) Agro-ecological zoning, guidelines, Food and Agricultural Organization of the United Nations, Rome

FAO (1996) Agro-ecological zoning, guidelines. Food and Agricultural Organization of the United Nations, Rome

FAO (2002) Conservation agriculture – optimizing soil moisture for palnt production; the significance of soil porosity. In FAO Soils Bulletin No. 79. FAO, Rome. ISBN 92-5-104625 5. 69 p

FAO (Food and Agriculture Organization) (2013) Climate smart agriculture source book main report E-ISBN 978-92-5-107721-4 (PDF). https://www.fao.org/climatechange/climatesmart Climate-Smart@fao.org

FAO (Food and Agriculture Organization) (2017) Review of the available Remote sensing tools, products, methodologies and data to improve crop production forecasts, Rome 2017, https://www.fao.org/publications. ISBN 978-92-5-109840-0

Fischer G, Shah M, Van H, Nachtergaele F (2006) Agro-ecological zones assessment

Fischer G, Nachtergaele F, Prieler S, Velthuizen HT, van Verelst L, Wiberg D (2008) Global Agro-ecological zones: model documentation. Food Agri Organ UN

Fiwa L, Vanuytrecht E, Wiyo KA, Raes D (2014) Effect of rainfall variability on the length of the crop growing period over the past three decades in central Malawi. Clim Res 62:45–58. https://doi.org/10.3354/cr01263

Gatot IS, Duchesne J, Forest F, et al (2001) Rainfall-runoff harvesting for controlling erosion and sustaining upland agriculture development. pp 431–439

Gregory PJ, Ingram JSI, Brklacich M (2005) Climate change and food security. Philos Trans R Soc B Biol Sci 360(1463):2139–2148. https://doi.org/10.1098/rstb.2005.1745

Guo HD, Zhang L, Zhu LW (2015) Earth observation big data for climate change research. Adv Clim Chang Res 6:108–117. https://doi.org/10.1016/j.accre.2015.09.007

Hastenrath S (2001) Variations of east African climate during the past two centuries. Clim Chang 50:209–217. https://doi.org/10.1023/A:1010678111442

Hengl T, AghaKouchak A, Tadic MP (2010) Methods and data sources for spatial prediction of rainfall. In Testik FY, Gebremichael M (eds) Rain-fall: state of the science. American Geophysical Union (AGU), pp 186–214 New Age International Publisher.

Henricksen BJ (1986) Determination of agro-ecological zones in Africa: ILCA activities and expectations. ILCA Bull 23:15–22

Herrero M, Ringler C, Steeg J, Van De, Koo J, Notenbaert A (2010) Climate variability and climate change and their impacts on Kenya's agricultural sector. ILRI, Nairobi, Kenya 1–56. https://doi.org/10.5539/jsd.v6n2p9

Holmes TRH, Owe M, De Jeu RAM, Kooi H (2008) Estimating the soil temperature profile from a single depth observation: a simple empirical heat flow solution. Water Resour Res 44:1–11. https://doi.org/10.1029/2007WR005994

IIASA/FAO, 2012. Global agro-ecological zones (GAEZ v3.0). IIASA/FAO, Laxenburg/Rome

Jiitzold R, Kutsch H (2000) Agro-ecological zones of the tropics, with a sample from Kenya. April 1982

Kamara SI, Jackson IJ (1997) A new soil-moisture based classification of raindays and dry days and its application to Sierra Leone. Theor Appl Climatol 56(3–4):199–213. https://doi.org/10.1007/BF00866427

Kenawy AM, Hereher ME, Robaa SM (2019) An assessment of the accuracy of MODIS land surface temperature over Egypt using ground-based measurements. Remote Sens 11. https://doi.org/10.3390/rs11202369

King'uyu SM, Ogallo LA, Anyamba EK (2000) Recent trends of minimum and maximum surface temperatures over eastern Africa. J Clim 13:2876–2886. https://doi.org/10.1175/1520-0442(2000)013<2876:RTOMAM>2.0.CO;2

Kumar DK, Reshmidevi TV (2013) Remote sensing applications in water resources. J Indian Inst Sci 93(2):163–187

Le Page Y, Vasconcelos M, Palminha A, Melo IQ, Pereira JMC (2017) An operational approach to high resolution agro-ecological zoning in West-Africa. PLo SONE 12(9): e0183737. https://doi.org/10.1371/journal.pone.0183737

Liou YA, Kar SK (2014) Evapotranspiration estimation with remote sensing and various surface energy balance algorithms-a review. Energies 7(5):2821–2849. https://doi.org/10.3390/en7052821

Managua C (2011) Future climate scenarios for Uganda Tea growing areas. Managa, CIAT

Marshall M, Funk C (2014) Examining evapotranspiration trends in Africa Examining evapotranspiration trends in Africa. https://doi.org/10.1007/s00382-012-1299-y

McCulloch JSG (1965) Tables for the rapid computation of the Penman estimate of evaporation. E Afr Agric J 30:286

Mendelsohn R (2008) The impact of climate change on agriculture in developing countries. J Nat Resour Policy Res 1:5–19. https://doi.org/10.1080/19390450802495882

Meneghini R, Eckerman J, Atlas D (1983) Determination of rain rate from a spaceborne radar using measurements of total attenuation. IEEE Trans Geosci Remote Sens 1:34–43

Mungai C (2017) Adoption and dissemination pathways for climate-smart agriculture technologies and practices for climate-resilient livelihoods in lushoto, Northeast Tanzania. Climate 5:63. https://doi.org/10.3390/cli5030063

Mutiga JK, Su Z, Woldai T (2013) Corrigendum to "using satellite remote sensing to assess evapotranspiration: case study of the upper Ewaso Ng'iro North Basin, Kenya". Int J Appl Earth Obs Geoinf 23(1):411. https://doi.org/10.1016/j.jag.2012.10.011

Ndirangu S, Mbogoh S, Mbatia O (2017) Effects of land fragmentation on food security in three agro-ecological zones of embu county in Kenya. Asian J Agri Ext Econ Soc 18(4):1–9. https://doi.org/10.9734/ajaees/2017/34321

Nsubuga FW, Rautenbach H (2018) Climate change and variability: a review of what is known and ought to be known for Uganda. Int J Clim Chang Strateg Manag 10(5):752–771. https://doi.org/10.1108/IJCCSM-04-2017-0090

Nsubuga FWN, Botai OJ, Olwoch JM, Rautenbach CJd W, Bevis Y, Adetunji AO (2014) La nature des précipitations dans les principaux sous-bassins de l'Ouganda. Hydrolog Sci J 59(2):278–299. https://doi.org/10.1080/02626667.2013.804188

Ochieng J, Kirimi L, Mathenge M (2015). Effects of climate variability and change on agricultural production: the case of small scale farmers in Kenya. NJAS Wagen J Life Sci 77:71–78. https://doi.org/10.1016/j.njas.2016.03.005. (2016)

Omondi PA o, Awange JL, Forootan E et al (2014) Changes in temperature and precipitation extremes over the Greater Horn of Africa region from 1961 to 2010. Int J Climatol 34:1262–1277. https://doi.org/10.1002/joc.3763
Osińska-Skotak K (2007) Studies of soil temperature on the basis of satellite data. Int Agrophys 21 (3):275–284
Onwuka B (2018) Effects of soil temperature on some soil properties and plant growth. Adv Plants Agric Res 8:34–37. https://doi.org/10.15406/apar.2018.08.00288
Parry M, Rosenzweig C, Iglesias A et al (1994) Climate change and world food security: a new assessment. Glob Environ Chang 9. https://doi.org/10.1016/S0959-3780(99)00018-7
Pearce D (2000) Policy frameworks for the ancillary benefits of climate change. Assess Ancillary Benefits Costs Greenh Gas Mitig 517–560
Pellarin T, Román-Cascón C, Baron C et al (2020) The precipitation inferred from soil moisture (PrISM) near real-time rainfall product: evaluation and comparison. Remote Sens 12:481. https://doi.org/10.3390/rs12030481
Pinheiro ACT, Privette JL, Guillevic P (2006) Modeling the observed angular anisotropy of land surface temperature in a savanna. IEEE Trans Geosci Remote Sens 44:1036–1047
Rowland J, Wood E, Tieszen LL, Lance K, Khamala E, Siwela B, Adoum A, Brown M (2007) Review of remote sensing needs and applications in Africa prepared by : contributors : development 1–124. https://doi.org/10.13140/RG.2.1.1101.3849
Sabri NSA, Zakaria Z, Mohamad SE, Jaafar AB, Hara H (2018) Importance of soil temperature for the growth of temperate crops under a tropical climate and functional role of soil microbial diversity. Microb Environ 33(2):144–150. https://doi.org/10.1264/jsme2.ME17181
Schneider U, Becker A, Ziese M, Rudolf B (2018) Global precipitation analysis products of the GPCC. Glob Precipitation Climatology Cent 1–14
Schollaert Uz S, Ruane AC, Duncan BN et al (2019) Earth observations and integrative models in support of food and water security. Remote Sens Earth Syst Sci 2:18–38. https://doi.org/10.1007/s41976-019-0008-6
Schreck CJ, Semazzi FHM (2004) Variability of the recent climate of eastern Africa. Int J Climatol 24(6):681–701. https://doi.org/10.1002/joc.1019
Seo SN, Mendelsohn R, Dinar A et al (2009) A Ricardian analysis of the distribution of climate change impacts on agriculture across agro-ecological zones in Africa. Environ Resour Econ 43:313–332. https://doi.org/10.1007/s10640-009-9270-z
Skawina T, Kossowski J, Stêpniewski W, Walczak R (1999) Physical properties of soils (in polish). In: Zawadzki S (ed) Soil science. PWRiL Press, Warsaw
Skolnik MI (1962) Introduction to radar. Radar Handbook, 2
Slatyer RO, Mabbutt JA (1964) Hydrology of arid and semiarid regions. Section 24 V.T
Stigter K, Walker S, Das HP, Huda S, Dawei Z, Jing L, Chunqiang L, Hurtado IHD, Mohammed AE, Abdalla AT, Bakheit NI, Al-Amin NKN, Yurong W, Kinama JM, Nanja D, Haasbroek PD (2010) Meeting farmers' needs for agrometeorological services: an overview and case studies
Sun Y (2009) Retrieval and application of land surface temperature. Geo-UtexasEdu 1:1–27
Thies B, Nauß T, Bendix J (2008) Precipitation process and rainfall intensity differentiation using meteosat second generation spinning enhanced visible and infrared imager data. J Geophys Res Atmosp 113:1–19. https://doi.org/10.1029/2008JD010464
Ullah W, Wang G, Ali G et al (2019) Comparing multiple precipitation products against in-situ observations over different climate regions of Pakistan. Remote Sens 11:628. https://doi.org/10.3390/rs11060628
Vlassova L, Perez-Cabello F, Nieto H et al (2014) Assessment of methods for land surface temperature retrieval from landsat-5 TM images applicable to multi-scale tree-grass ecosystem modeling. Remote Sens 6:4345–4368. https://doi.org/10.3390/rs6054345
Vrieling A, De Leeuw J, Said MY (2013) Length of growing period over Africa: variability and trends from 30 years of NDVI time series. Remote Sens 5:982–1000. https://doi.org/10.3390/rs5020982
Wango TJL, Musiega D, Mundia CN (2018) Assessing the suitability of the worldclim dataset for ecological studies in Southern Kenya. J Geogr Inform Syst 10(06):643–658. https://doi.org/10.4236/jgis.2018.106033

White MA, de Beurs KM, Didan K, Inouye DW, Richardson AD, Jensen OP, O'Keefe J, Zhang G, Nemani RR, van Leeuwen WJD et al (2009) Intercomparison, interpretation, and assessment of spring phenology in North America estimated from remote sensing for 1982–2006. Glob Chang Biol 15:2335–2359

Zuma-Netshiukhwi GN, Stigter KC, Walker S (2016) Improving agricultural decision making using weather and climate information for farmers, south-western Free State, South Africa. Net J Agric Sci 4:67–77

3

Hydrological Dynamics Assessment of Basin Upstream–Downstream Linkages Under Seasonal Climate Variability

Oseni Taiwo Amoo, Hammed Olabode Ojugbele, Abdultaofeek Abayomi and Pushpendra Kumar Singh

Contents

O. T. Amoo (✉)
Risk and Vulnerabilty Science Centre, Walter Sasilu University , Eastern Cape, South Africa
e-mail: ejire36@gmail.com

H. O. Ojugbele
Regional and Local Economic Development Initiative, University of KwaZulu-Natal, Westville, South Africa

A. Abayomi
Department of Information and Communication Technology, Mangosuthu University of Technology, Umlazi, Durban, South Africa

P. K. Singh
Water Resources Systems Division, National Institute of Hydrology, Roorkee, India

Abstract

The impacts of climate change are already being felt, not only in terms of increase in temperature but also in respect of inadequate water availability. The Mkomazi River Basins (MRB) of the KwaZulu-Natal region, South Africa serves as major source of water and thus a mainstay of livelihood for millions of people living downstream. It is in this context that the study investigates water flows abstraction from headwaters to floodplains and how the water resources are been impacted by seasonal climate variability. Artificial Neural Network (ANN) pattern classifier was utilized for the seasonal classification and subsequence hydrological flow regime prediction between the upstream–downstream anomalies. The ANN input hydroclimatic data analysis results covering the period 2008–2015 provides a likelihood forecast of high, near-median, or low streamflow. The results show that monthly mean water yield range is 28.6–36.0 m^3/s over the Basin with a coefficient of correlation (CC) values of 0.75 at the validation stage. The yearly flow regime exhibits considerable changes with different magnitudes and patterns of increase and decrease in the climatic variables. No doubt, added activities and processes such as land-use change and managerial policies in upstream areas affect the spatial and temporal distribution of available water resources to downstream regions. The study has evolved an artificial neuron system thinking from conjunctive streamflow prediction toward sustainable water allocation planning for medium- and long-term purposes.

Keywords

Seasonal classifier · Climate variability · Sustainable water allocation · Artificial neuron network · System thinking

Introduction

Background

The seasonal hydrological flow regime is of utmost importance in understanding potential water allocation schemes and subsequent environmental standards flow regulation. Streamflow is a fundamental component of the water cycle and a major source of freshwater availability for humans, animals, plants, and natural ecosystems. It is severely being impacted upon by human activities and climate change variability (Makkeasorn et al. 2008; Null et al. 2010).

Climate change variability has a profound impact on water resources, biophysical, and socioeconomic systems as they are highly interconnected in complex ways (Graham et al. 2011; Null and Prudencio 2016). A change in any one of these induces a change in another. The effects of climate indices on streamflow predictability are seasonal and region dependent (Katz et al. 2002). Although, there are different approaches to assess a basin response to climate variables change. Hydrologists have devised different approaches to investigate how water, the environment, and human activities are mutually dependent and interactive under various climatic conditions (Bayazit 2015). Most of the morphoclimatic challenge studies in any region are usually customized and basin specific. Though many studies have investigated the effects of changing temperature, precipitation, and evaporation as climate variables on water resources management (Katz 2013; Kundzewicz et al. 2008; Null and Prudencio 2016; Taylor et al. 2013), only few have examined seasonal fluctuations of water allocations in nonstationary climates and captured the hydrologic variability trend characterization based on different temporal scales on the whole catchment (Egüen et al. 2016; Gober 2018; Katz 2013; Poff 2018).

Topographic effects on rainfall vary seasonally due to the particular hydrologic process in the region which makes seasonal water allocation and varying environmental flow projection to be based on available meteorological past data for various design purposes (Jakob 2013). Seasonal fluctuations are commonly observed in quarterly or monthly hydrologic flow regime studies. As seasonality is a dominant feature in time series (Sultan and Janicot 2000), hydrologists have developed methodologies to routinely deseasonalized data for modelling and forecasting different annual conditions (Benkachcha et al. 2013). The common variation of inflow from one season to another mainly reflects the climatic variability which includes seasonality of rainfall and amount of evapotranspiration which is dependent on air temperature as well as precipitation in the basin (Ufoegbune et al. 2011). The understanding of these hydrological dynamics in a basin is crucial for sustainable water allocation planning and management. Different literature work reviews across the globe have found a nonlinear or linear relationship between station elevation and rainfall pattern, e.g., the Kruger National Park, South Africa and Mount Kenya, East Africa (Hawinkel et al. 2016; MacFadyen et al. 2018). Likewise, a linear pattern was found in a Spatio-temporal Island study in a European City (Arnds et al. 2017; Sohrabi et al. 2017) while a nonlinear relation exists in a central Asia Basin study (Dixon and Wilby 2019).

Rainfall heterogeneity obviously needs to be considered in a number of hydrological process studies in larger catchments area since it influences: infiltration dynamics, hydrograph regime, runoff volume, and peak flow (Bonaccorso et al. 2017; Fanelli et al. 2017; Gao et al. 2019; Tarasova et al. 2018). However, some hydrology studies still rely on small numbers of synoptic-scale rainfall measurements, and the problem of limited rainfall gauges is common in many watershed investigations, especially in developing countries (Birhanu et al. 2019; Liu et al. 2017; Moges et al. 2018). Therefore, prior to investigating the watershed functions and their contributions to the Mkomazi River flow, we examined the spatial attributes of rainfall in the area based on existing data. The past works conducted on the Basin includes Flügel et al.'s (2003) that had used a geographical information system (GIS) to regressed the local rainfall to the elevation, while Oyebode et al. (2014) has used genetic algorithm and ANN for the streamflow modelling at the upstream. Taylor et al. (2003) and Wotling et al. (2000) used rainfall intensity distribution and principal component analysis (PCA) to assess the complexity of the terrain in addition to the elevation. In general, differences in rainfall patterns may have involved a combination of two statistical outcomes: a shift in the mean and a change in the scale of the distribution functions. The gamma distribution is a popular choice for fitting probability distributions to rainfall totals because its shape is similar to that of the histogram of rainfall data (Kim et al. 2019; Svensson et al. 2017).

Similarly, Najafi and Kermani (2017) observed that in recent years, many researchers have used various empirical rainfall-runoff models to study the impacts of climatic change on basin hydrology. However, a good understanding of the future rainfall distribution across these zones is of vital importance if any meaningful development is to take place in the water resource management and agricultural sector which has been of utmost priority in recent times (Arnds et al. 2017). However, in order to address the foregoing issues, there is a need for the study of this nature. It will help in developing long-term strategic plans for climate change adaptation and mitigation measures and implementing effective policies for sustainable water resources and management of irrigation projects and reservoir operations for the overall sustenance of human well-being in the region (Al-Kalbani et al. 2014). This chapter utilizes regression analysis to investigate the upstream–downstream linkages under seasonal climate variability. Hydrological trend characterization was based on available morphoclimatic past data. Sen's slope (Pettitt's) abrupt change detection and the Mann-Kendall parametric trend analysis were used in detecting long-term variability in precipitation while ANN was used for seasonal classifiers and potential future streamflow quantification. Quintile regression was used to establish the relationship between climate indicators (historical rainfall and streamflow) and past catchment conditions to forecast future hydrological dynamics in the MRB. The main novelty in this study is that such a time-series representation is useful for considering the influence of projected shifts in environmental factors on the hydrologic budget, and subsequent coping strategies can be provided.

The Mkomazi catchment is located in the U Basin within the semiarid province of KwaZulu- Natal in South Africa as shown in Fig. 1. It is the third largest catchment in the province, draining an area of about 4400 km^2 with several large tributaries like Loteni, Nzinga, Mkomanzi, and Elands Rivers. The climatic condition of the study area varies with the seasonality of dry winters and wet summers (Flügel et al. 2003). Rainfall distribution is inconsistent along the catchment, ranging from nearly 1200 mm per annum at the headwaters to 1000 mm p.a. in the middle and 700 mm p.a. in the lower reaches of the catchment with highly intra- and interseasonal streamflows (Flügel and Märker 2003; Oyebode et al. 2014; Taylor et al. 2003). Prior water allocations were entirely based on "who got there first," which have become unstable and irrational with climate change. Thus, examining how climate-driven spatial and temporal changes to streamflows may reallocate water among the riparian users considering its seasonal variability is of immense importance. The Mkomazi River can be subdivided into five physiographic zones as shown in Fig. 2, namely:

(I) The coastal lowlands up to 620 m (mean annual sea level – m a.s.l.)
(II) The interior lowland area ("middle berg area") from 620 to 1079 m (m.a.s.l.)

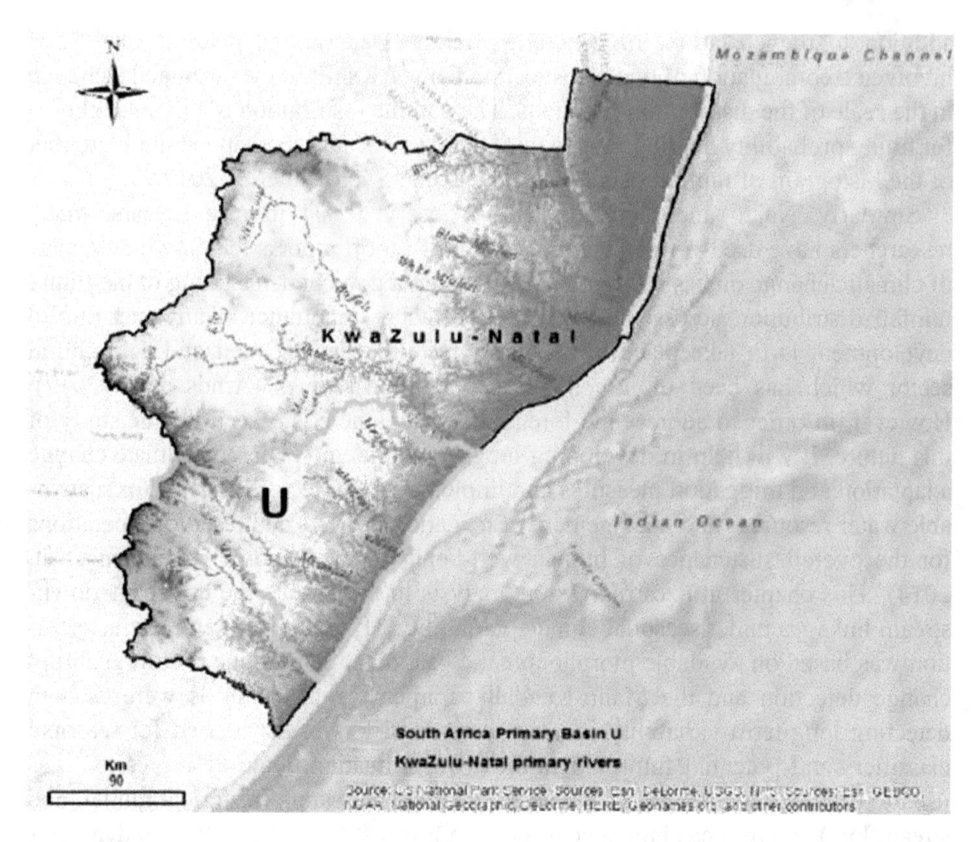

Fig. 1 The study area: Mkomazi River basin

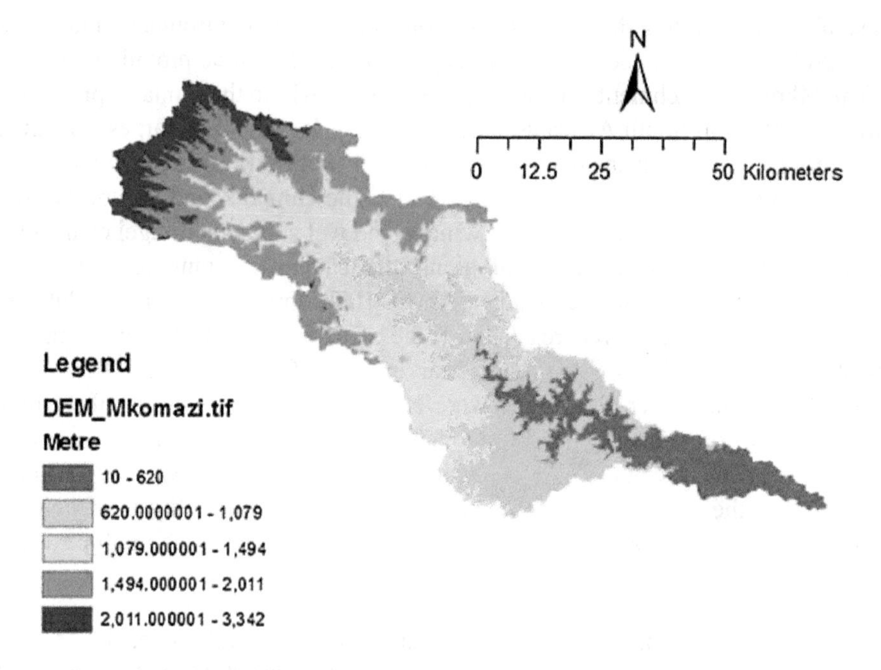

Fig. 2 Mkomazi Physiographic description with elevation zones

(III) Lowland area up to 1494 m (m.a.s.l.)
(IV) The mountain area up to 2011 m (m.a.s.l.)
(V) The highlands, with elevations up to 3342 m (m.a.s.l.)

Like other catchments in Southern Africa, the study area is characterized by high varying seasonality of dry winters and wet summer (Schulze and Pike 2004a). Many authors have suggested longer record lengths, approximately 20-year oscillations interval, in measuring the seasonal hydrological fluctuation (Abhishek et al. 2012; Schulze 1995; Tyson 1986). This variability was adjudged to be as a result of the influence of the currents in the Atlantic and the Indian Ocean that surround the country. The cold Benguela Current in the Atlantic Ocean (South West) brings not only cold air but also influences the pressure system (Haigh et al. 2010; King et al. 2000), while the warmer currents of the Indian Ocean influence the milder warmer sea temperatures and the humid air on the Northeastern Coastline in the country. These seasons are opposite to the seasons in the northern hemisphere (King et al. 2000). There are no clear-cut seasonal calendar based on phenology for the country; however, conventional seasonal variability run through summer (Dec–Feb), autumn (March–May), winter (June–August), and spring usually between September and November, to which the average temperature ranges from 20 to 30; 10 to 15, 7 to 10, and 15 to 20 °C respectively (De Coning 2006; Schulze and Pike 2004a, b).

Procedural Summary

The various stations source data from South Africa Weather Information System (SAWs), the Agricultural Research Council (ARC), and the Department of Water Affairs (DWA), South Africa, were processed and subjected to a rigorous scientific method to test their accuracy, reliability, homogeneity, consistency, and localization gaps. The detection of trends in a series of extreme values needs highly reliable data. Thus, only five stations met the above requirements and, as a result, corrected data sets were available for the hydrological years from 2008 to 2015 (for U10L 30530, U10J 30587, U10L 30813, U10 at Shelburn, and U10 at Giant Castle locations). Table 1 gives the statistical summary of the selected stations variables for (1985-2015) years.

Since the studied variables have different variances and units of measurements as shown in Table 1, the data set was standardized. This step was done by subtracting off the mean and dividing by the standard deviation (Ikudayisi and Adeyemo 2016). At the end of the standardization process, each variable in the dataset is converted into a new variable with zero mean and unit standard deviation. The original and standardized variables are displayed in Figs. 3 and 4, respectively. The standardized results are needed for minimization of bias and accumulation of predicted error from the observed data. These data were further subjected to various test/processing regarding homogeneity, consistency, and gaps closure before adaptation for model inputs. This helps to improve their predictive abilities and reduce uncertainty in data usage.

Table 1 Statistical summary of the selected stations year (1985–2015) variables

Variables	Unit	Minimum	Maximum	Mean	Std. Dev
MaxT	°C	14.400	33.170	24.408	3.442
MinT	°C	−5.000	20.620	10.392	4.942
Solar	MJ/m^2	0.030	36.450	15.691	3.884
windsp	[m/s]	0.650	3.367	1.794	0.561
MaxRH	%	33.000	99.930	78.425	16.712
MinRH	%	3.000	66.340	32.765	15.144
R Evap	mm	9.000	194.550	97.307	30.259
Rain	mm	0.000	353.200	64.394	61.918
Runoff	m^3/s	2.018	123.639	26.217	26.216

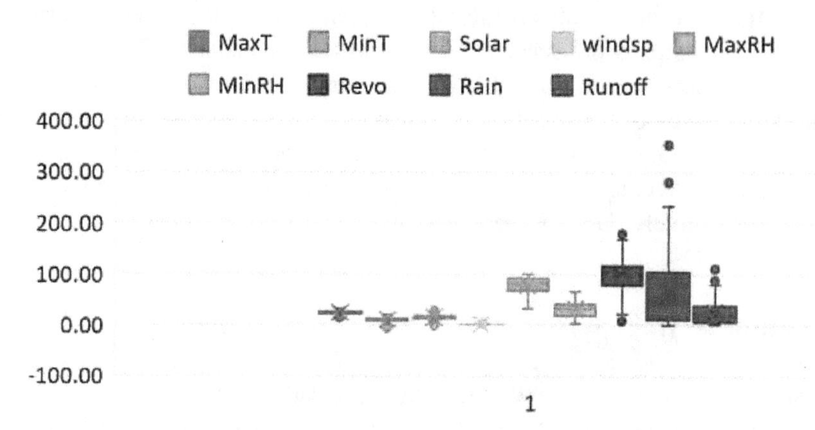

Fig. 3 Original data distribution of the climatic variables

Thus, the normalized data were used as input to the ANN classifier for seasonal forecasting and classification. Microsoft Statistical software XLSTAT by Addinsoft was used for Factor Analysis (FA) to explain the contribution of the unobserved common features in a target event from observed ones. FA as a choice of principal component analysis (PCA) was employed to reduce the variety of hydroclimatic data matrix to form a few selected derived component variables, which form a true representative of the original sets. The FA relates the trend, pattern, and fluctuations into wet or dry season in order to identify hydroclimatic variables responsible for streamflow characteristics as a basis for determining available water. This explains seasonal variation in water availability in the downstream environments. FA shows their level of influence and degree of different percentage contribution to the total streamflow volume (latent class). The regression relationships between the collected hydroclimatic data were developed as scatter plots and correlated to know their significance parameter sensitivity. Thereafter, the seasonal variability and trend detection were evaluated using Sen's method abrupt change detection and Mann-Kendall trend analysis in forecasting the hydrologic flow regime.

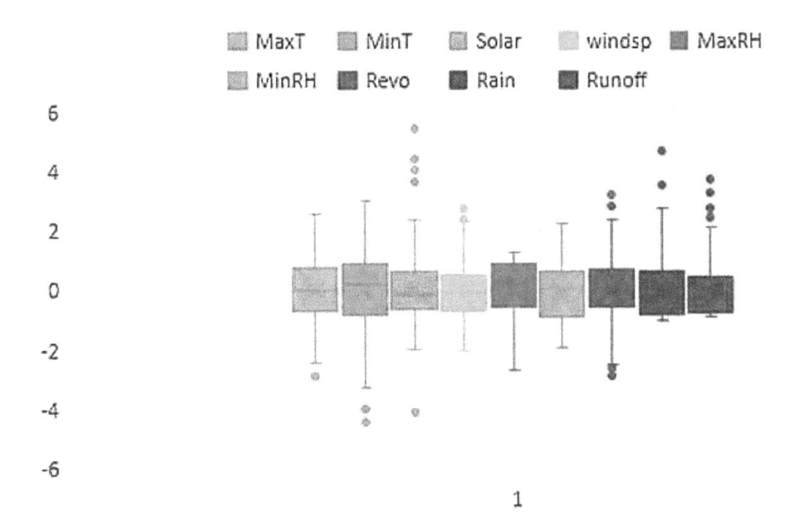

Fig. 4 Data standardization (normalization) of the climatic variables data

Key Findings

Seasonal Trend and Variability Changes among the Climatic Variables

A cursory run of the nonparametric and parametric approaches of Mann–Kendall and Sen's methods across the months (Dec–Nov) for the study duration (1985–2015) shows the discernible trend and variability changes among the climatic variables. The value of Sen's slope is given in Table 2 column 7. The closer it is to 0, the lesser the trend, while the sign of the slope indicates if the trend is increasing or decreasing. A Mann–Kendall test with a very high positive value of S (column 4) is an indication of an increasing trend while a very low negative value indicates a decreasing trend.

As shown in Table 2, decreasing trends are found generally among the variables across the months except for maximum temperature and solar radiation. This is understandable, as increased solar radiation brings about an increase in temperature. Increases in temperature and radiation forcing variables also alter the hydrological cycle. The resultant effect determines the amount of precipitation, its frequency, intensity duration, and the type of rainfall. In all, the hydro-meteorological variable shows a decreasing trend except for maximum temperature and evapotranspiration in the seasonal distribution pattern. The results show how changes in temperature and evapotranspiration could affect both the timing and total amounts of runoff, though the patterns of possible changes are both spatially and temporally complex. Future changes to allocation of water during seasonal water shortages is an important decision, which not only needs to be better coordinated within any given legal jurisdiction but needs to be better coordinated across any upstream and downstream uses and users.

Table 2 The results of the Sen's slope and Mann–Kendall test. Seasonal Mann–Kendall Test/period = 12/serial independence /two-tailed test

Variables [1]	Unit [2]	Kendall's tau [3]	S [4]	Var(S) [5]	p-value (Two-tailed) [6]	Sen's slope [7]	Trend [8]
MaxT	°C	0.009	34.000	0.824	0.05	−0.026	Increasing
MinT	°C	−0.053	−191.000	0.200	0.05	−0.037	Decreasing
Solar	MJ/m^2	−0.083	−296.000	0.046	0.05	−0.026	Decreasing
Windsp	[m/s]	−0.032	−115.000	0.442	0.05	−0.004	Decreasing
MaxRH	%	−0.205	−737.000	<0.0001	0.05	−0.546	Decreasing
MinRH	%	−0.168	−604.000	<0.0001	0.05	−0.475	Decreasing
R Evap	mm	0.067	242.000	0.104	0.05	0.318	Increasing
Rain	mm	−0.166	−596.000	<0.0001	0.05	−0.408	Decreasing
Runoff	m^3/s	−0.039	−136.000	0.360	0.05	0	Decreasing

Factor Analysis (FA) Findings

FA explores the dimensionality of a measurement instrument by finding the smallest number of interpretable factors needed to explain the correlations among the set of variables. This was particularly useful in the analysis of the meteorological parameters input for the ANN model. It places no structure on the linear relationships between the observed variables and the factors but only specifies the number of latent factors, determines the quality of the measurement instrument, identifies variables that are poor factor indicators, and are also poorly measured (Hu et al. 2014). The FA loading is shown in Table 3 to reduce the data into a smaller number of components which indicates what constructs underlies the data latent class. The bold squared cosine values depict the most significant variables that affect discharge flow.

The diagrammatical representation of the Factor Analysis as shown in Fig. 5 indicates each variable's level of significance on how they contribute to the total streamflow volume latent class. The table shows their proportionate percentage significance towards streamflow. All the meteorological data influence the streamflow except the wind speed presented in the statistical factor analysis biplot which stands alone thus depicting the least effect. This may suggest that air temperature is a more important climatic factor for water mass balance than precipitation.

ANN Pattern Classifier and Flow Regime Variation

Insights into the ANN configuration for the four seasons' classification are given in Fig. 6. The ANN internal algorithm using the gradient descent method for the hidden layer was able to classify them into the four prominent seasons based on collected data which were sorted on monthly basis and labeled inclusively with 1 - Summer, 2- Autumn, 3-Winter, and 4- Spring, respectively. The developed ANN model in the present study consisted of eight input layers that represent the input vectors of the hydro-meteorological parameters considered and four output layers representing

Table 3 Factor loadings of the variables

	Factor Loadings								
Variables	F1	F2	F3	F4	F5	F6	F7	F8	F9
MaxT	**0.629**	0.004	0.235	0.038	0.008	0.046	0.022	0.006	0.013
MinT	**0.869**	0.006	0.000	0.036	0.016	0.011	0.026	0.016	0.020
Solar	**0.606**	0.186	0.009	0.003	0.037	0.033	0.122	0.004	0.000
Windsp	0.027	**0.372**	0.369	0.142	0.061	0.019	0.000	0.010	0.000
MaxRH	0.419	**0.441**	0.013	0.036	0.018	0.003	0.011	0.057	0.002
MinRH	**0.552**	0.169	0.150	0.045	0.009	0.035	0.004	0.026	0.010
Revo	0.230	**0.498**	0.028	0.041	0.049	0.147	0.005	0.002	0.000
Rain	**0.377**	0.014	0.236	0.204	0.122	0.011	0.034	0.001	0.001
Runoff	**0.440**	0.027	0.000	0.247	0.257	0.020	0.003	0.006	0.000

Values in bold correspond to each variable to the factor for which the squared cosine is the largest

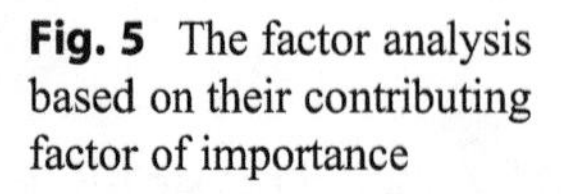
Fig. 5 The factor analysis based on their contributing factor of importance

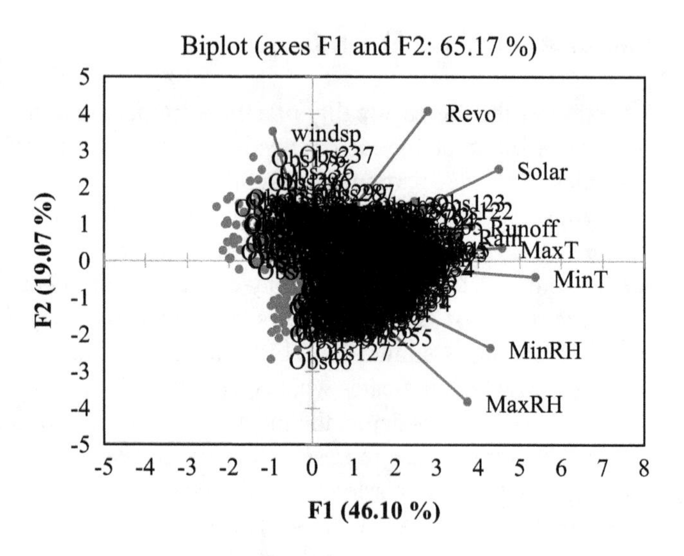

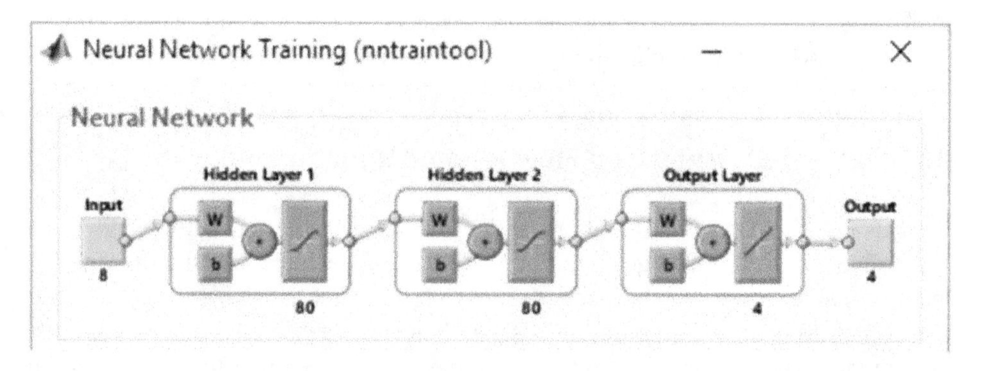

Fig. 6 MATLAB interphase for seasonal classification using hydro-meteorological data

the four seasons. The different monthly hydro-meteorological data of each of the six-catchment location serves as the raw data which was preprocessed to obtain the input vectors that was fed into the ANN. The MATLAB interface indicating the size of the input vector, a number of hidden layers, and neurons applied at the layers (which were experimentally determined) as well as the output layer is presented in Fig. 6, while Fig. 7 shows the confusion matrix obtained for the seasonal classification.

The confusion matrix indicating a 93.7% classification accuracy was achieved by the ANN classifier in cataloging the labeled data into the appropriate season. Given any unknown data set for the test, the developed model was able to distinguish it and classify it into the appropriate season. The corresponding number in the diagonal elements of the matrix indicates the number of instances that could be correctly classified for each of the seasons and how best the ANN classifier can recognize and distinguish data mined for each seasonal (summer, autumn, winter, and spring) classification respectively. The result of the optimal seasonal discharge forecast model is as presented in Fig. 8.

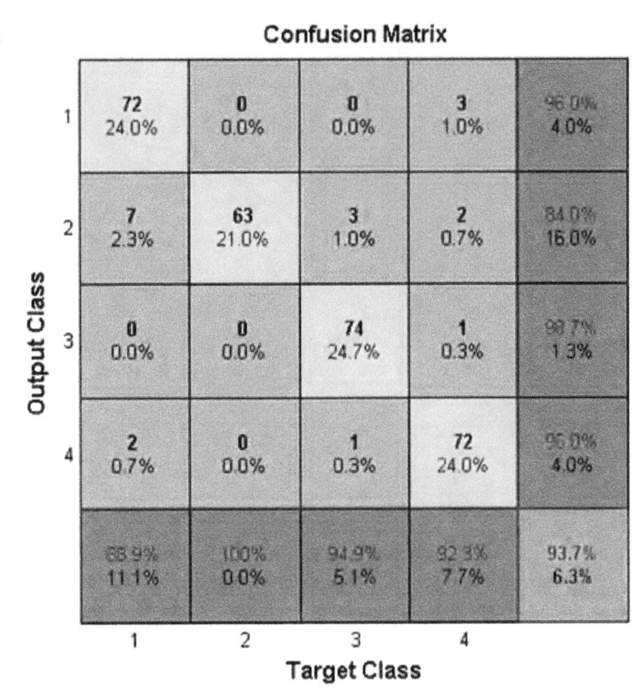

Fig. 7 Seasonal classification by ANN model

The ANN forecasted results provide the likelihood of high, near medium or low streamflow. The changes in the discharge regime were identified with the 5th, 10th, and 90th tercile streamflow magnitude curve as depicted in Fig. 9.

The streamflows in winter (wetter months) have increased slightly over the time period, whereas streamflows in summer (drier months) have decreased slightly. Figure 9 represents the estimated seasonal catchment yield of higher surface flow in winter with decreasing lower base flow across the seasons. Streamflow is observed to be at its lowest in the autumn period as the results further assert that climate change is real and have significance effects on the river flow regime (Archer et al. 2010; Schulze and Pike 2004a; Taylor et al. 2003; Viviroli et al. 2010). The coping strategy suggests sustainable use of any resource relies on the action of a number of regulatory mechanisms that prevent the user from reducing the ability of the system to provide services. "Polluter pays" principle was conceived as a way to proportionally allocate the effects of such alterations to those users that are responsible for them, thus producing a regulatory effect on the use of the resource. Also, the water allocation criteria include economic criteria with the aim of optimizing the economic value of the water resource.

Model Performance and Evaluation Measures

For improved accuracy, the root means square error (RMSE) and coefficient of correlation (CC) was used for performance evaluation measures during ANN

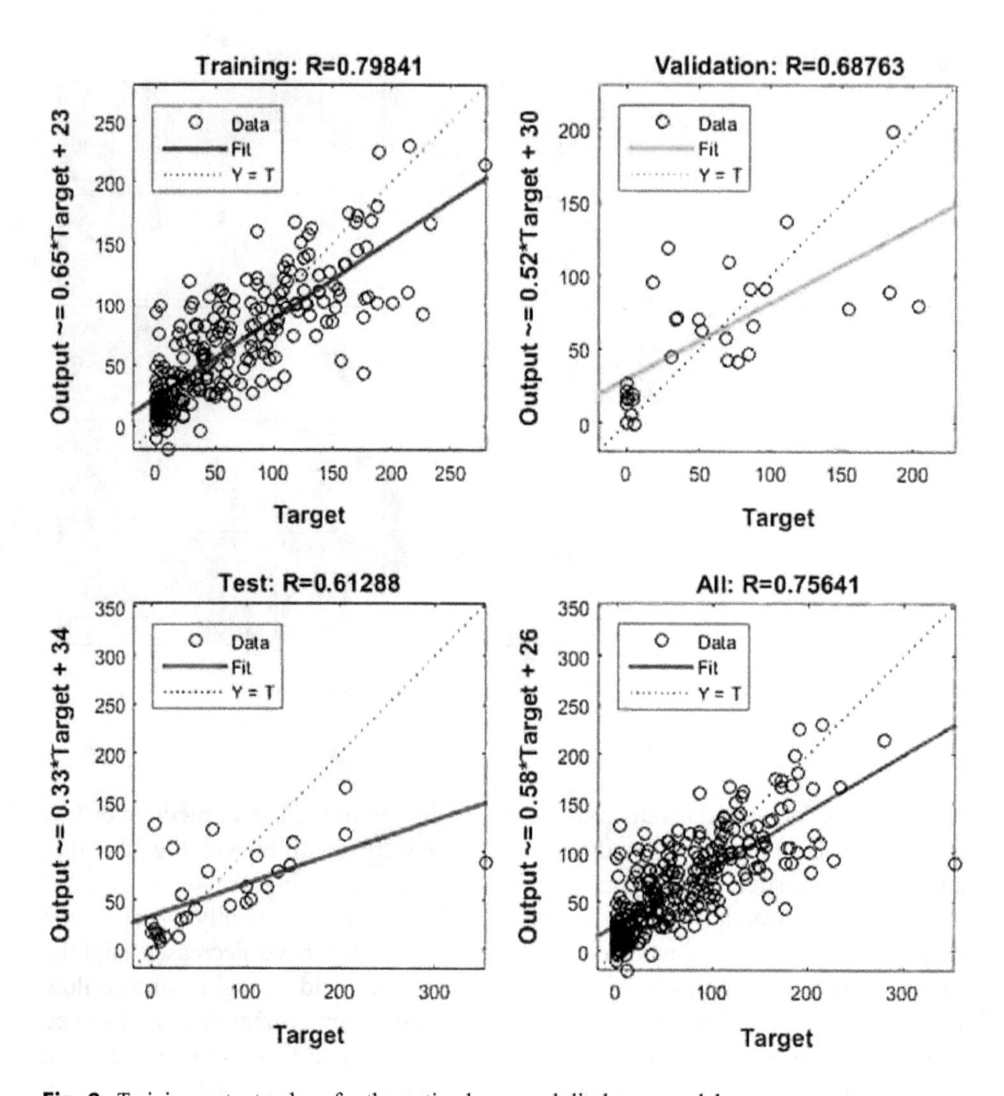

Fig. 8 Training output values for the optimal seasonal discharge model

training, testing, and validation procedures (Paswan and Begum 2014). They are defined as shown in Eqs. (1) and (2), respectively.

$$RMSE = \sqrt{\frac{1}{n}\sum_{i=1}^{n}(Q_i - P_i)^2} \quad (1)$$

$$CC = \frac{\sum_{i=1}^{n}\left[\left(Q_i - \widehat{Q}_i\right).\left(P_i - \widehat{P}_i\right)\right]}{\sqrt{\sum_{i=1}^{n}\left(Q_i - \widehat{Q}_i\right)^2.\sum_{i=1}^{n}\left(P_i - \widehat{P}_i\right)^2}} \quad (2)$$

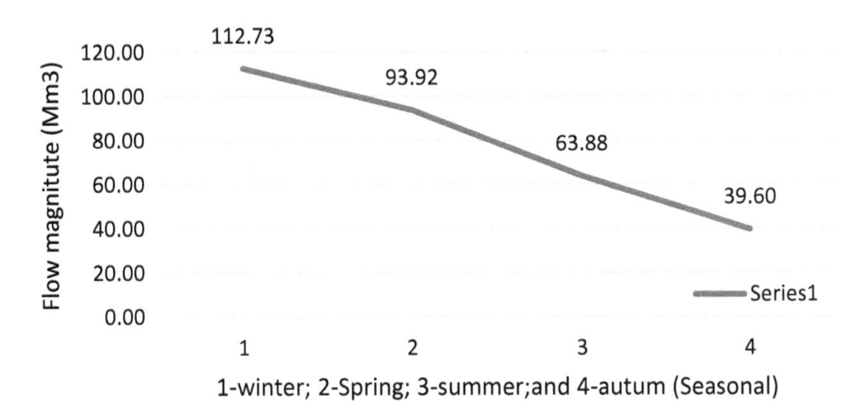

Fig. 9 Seasonal variability flow dynamic across the river

where Q_i is the observed value at time i, P_i is the simulated value at time i and $\bar{P}$ is the mean for the observed values. Observed runoff is used for model calibration and validation. Correlations are useful because they can indicate a predictive relationship that can be exploited in practice. The performance measures of RMSE and CC values obtained are 29% and 61% respectively during the calibration period. During the verification period, the RMSE and CC values are slightly improved as 18% and 75% were achieved respectively. Figure 10 shows the result of the training test as well as the optimum prediction performance of the network architecture.

Using FA as part of PCA has helped in screening the data and identifying the level of importance based on their contributing factor. The observed MRB runoff at (Mkomazi drift UIH009) station was used to compare with the ANN-model-simulated output in Shrestha (2016) Calibration Helper v1.0 Microsoft Excel Worksheet. The model performance was evaluated using statistical parameters and the result is represented in Fig. 11. Comparing ANN simulated value and observed runoff variables show a satisfactory ANN forecasting model for the seasons run.

The resulting data from the four seasons were detrended and deseasonalized (Wang et al. 2011) before forecasting the time series using neural networks. The results show that the neural networks with the right configurations give almost the same accuracy with or without decomposition of the time series. The streamflows in wetter months have increased slightly over the time period from 1985 to 2015, whereas streamflows in drier months have decreased slightly. These trends are also evident for the minimum and maximum temperature and relative humidity multiday events. The relationship between runoff and Mkomazi rainfall has been high and stable over the recording period. Other data analysis not presented shows a decreasing trend of precipitations. The summer rainfall variations are related more closely to maximum than minimum temperatures, with higher temperatures associated with lower rainfall. Lower rainfall in winter tends to be linked with higher maximum and lower minimum temperatures. These relationships were relatively stable over time. For this reason, there is a need to consider a range of possible future climate

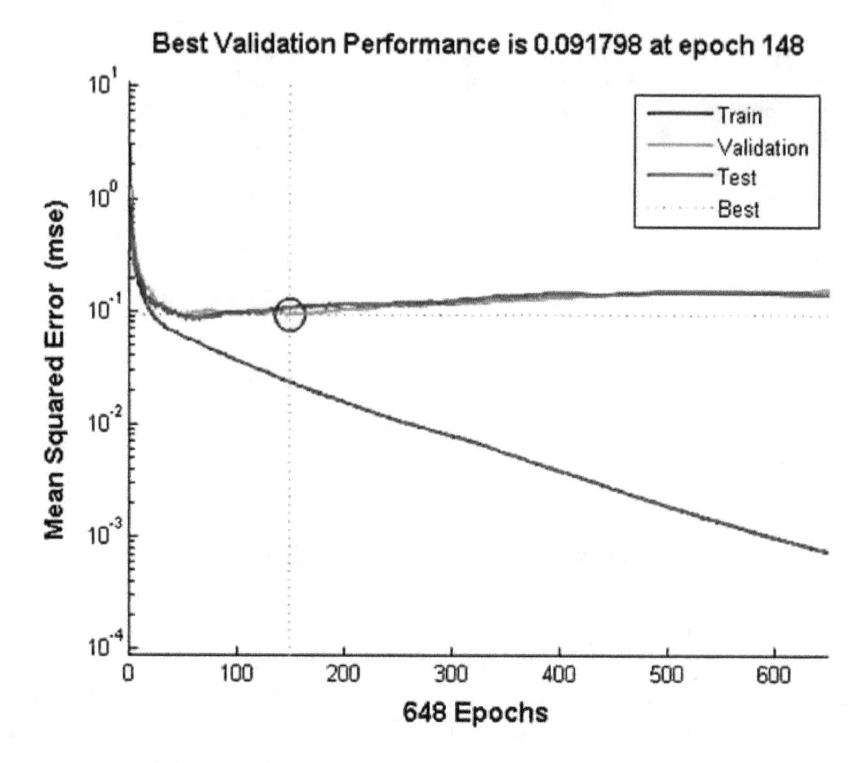

Fig. 10 Performance evaluation measures

conditions in a region of increasing population, coupled with increasing demand for water resources for domestic, agricultural, and industrial activities and how these will affect water availability.

Limitations of the Study

Like all modeling studies, this research has assumptions and uncertainties which limit the findings. Many of the previous studies indicate that stationary climatic and streamflow data were used while we applied the historical probability that each water year type is based on the 2008 water year. This method assumes stationary climatic water year types. Future water year type frequencies varied somewhat with climate change (nonstationarity in data), although those changes are statistically insignificant. Also, decadal to multidecadal variability which includes the probability of extreme floods and droughts (Herrfahrdt-Pähle 2010) were not considered. Our findings likely underestimate water allocation impacts from extreme floods and droughts anticipated with climate change. In addition, for environmental flow allocations and streamflow, there is still substantial room for model misspecification through overfitting, thus the selection of optimal internal neural algorithms choice again raises the issue of robust neural network modeling.

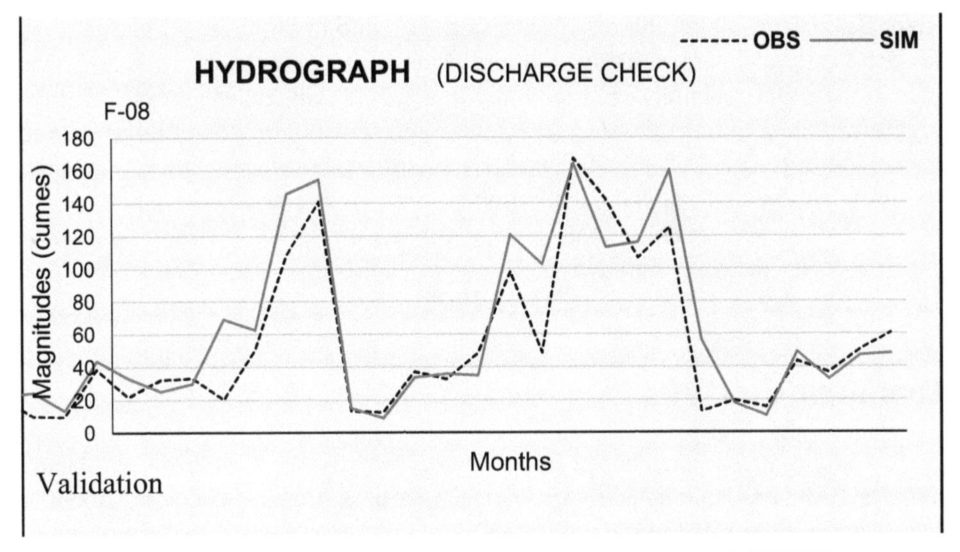

Output statistics	Calibration	Validation
NSE	0.62	0.77
NSE_{rel}	0.47	-0.32
R^2	0.77	0.83
wR^2	0.62	0.69
RSR	0.53	0.48
PBIAS	3.10%	13.90%

Fig. 11 Comparison of ANN simulated value and observed runoff

Seasonal Coping and Adaption Strategies to Climate Variability

Practical field investigation of views of stakeholders and their involvement should be simulated with model computer software in evolving solutions to water availability. Government political will and strategical alignment of various duplicating agencies in providing data for Sub-seasonal to Seasonal (S2S) Prediction should be harmonized toward a unified goal. Historically disadvantaged communities must be strongly supported and be encouraged to apply for water use licenses and the region must be in readiness to fast-track the processing of these water use license applications. Other available strategies to cope with climate variability include integrated water resources management by improving public water supply, regulating final users by facilitating the emergence of mediating agencies, the use of energy pricing and supply to manage agricultural water use overdraft, and better sensitization campaigns on rain water capture and recharge. The formalization of the water sector through the transfer of water rights and the more efficient use of water resources are among evolving strategies to achieve a sustainable water supply for the Mkomazi

River Basin. Turning all these adaption strategies into opportunity requires a need for both water technology innovation and water behavioral change to manage the scarcity of water resources in a sustainable manner. This chapter offers intuitive suggestions on how human being policies and cautious approaches can be used to manage and sustain the already depleted environment. Such intuitive agendum should be catalyzed through the institutionalization of proactive and capacity developmental platforms where climate change experts transfer knowledge, skills and expertise to upcoming researchers.

Conclusion

Understanding of climate change is continually improving, but the future climate remains uncertain (Yuan et al. 2016). For this purpose, a regression nonparametric approach consisting of the Mann–Kendall test and Sen's method and a parametric approach (ANN) based on factor analysis of extremes statistical theory have been applied. Owing to the seasonal character of the upstream–downstream variables linkages, the result concludes that all the variables considered the temperature as the most important factor in the estimation of streamflow. The results suggest a significant impact of input vector length, a few hidden nodes neurons, and the choice of activation function as ANN potentiality in characterizing the individual season and project maximum likelihood trend for the surface water patterns. Thus, careful assessment of the available water resources and reasonable needs of the basin/region in foreseeable future for various purposes must be based on reliable information concerning the meteorological variable trend which in turn impacts the peak flow to be expected after a rainstorm of a given probability of occurrence. The methods applied further confirms our assertion that these patterns are indeed unique across each month in the season. The water resources manager and agricultural water management sector would find optimal use in the developed ANN classifier model that links hydrologic variability on different temporal trend scales based on available past data and rainfall anomaly.

Acknowledgments The financial support from the National Research Foundation (NRF) of South Africa in research and innovation advancement – UID Grants: 103232 – is well appreciated.

References

Abhishek K, Singh M, Ghosh S, Anand A (2012) Weather forecasting model using artificial neural network. Procedia Technol 4:311–318

Al-Kalbani MS, Price MF, Abahussain A, Ahmed M, O'Higgins T (2014) Vulnerability assessment of environmental and climate change impacts on water resources in Al Jabal Al Akhdar, Sultanate of Oman. Water 6(10):3118–3135

Archer E, Engelbrecht F, Landman W, Le Roux A, Van Huyssteen E, Fatti C, . . . Colvin C (2010) South African risk and vulnerability atlas. Department of Science and Technology, South Africa

Arnds D, Böhner J, Bechtel B (2017) Spatio-temporal variance and meteorological drivers of the urban heat island in a European city. Theor Appl Climatol 128(1–2):43–61
Bayazit M (2015) Nonstationarity of hydrological records and recent trends in trend analysis: a state-of-the-art review. Environ Process 2(3):527–542
Benkachcha S, Benhra J, El Hassani H (2013) Causal method and time series forecasting model based on artificial neural network. Int J Comput Appl 75(7):37
Birhanu BZ, Traoré K, Gumma MK, Badolo F, Tabo R, Whitbread AM (2019) A watershed approach to managing rainfed agriculture in the semiarid region of southern Mali: integrated research on water and land use. Environ Dev Sustain 21(5):2459–2485
Bonaccorso B, Brigandì G, Aronica GT (2017) Combining regional rainfall frequency analysis and rainfall-runoff modelling to derive frequency distributions of peak flows in ungauged basins: a proposal for Sicily region (Italy). Adv Geosci 44:15
De Coning C (2006) Overview of the water policy process in South Africa. Water Policy 8(6): 505–528
Dixon SG, Wilby RL (2019) A seasonal forecasting procedure for reservoir inflows in Central Asia. River Res Appl 35(8):1141–1154
Egüen M, Aguilar C, Solari S, Losada M (2016) Non-stationary rainfall and natural flows modeling at the watershed scale. J Hydrol 538:767–782
Fanelli R, Prestegaard K, Palmer M (2017) Evaluation of infiltration-based stormwater management to restore hydrological processes in urban headwater streams. Hydrol Process 31(19): 3306–3319
Flügel W-A, Märker M (2003) The response units concept and its application for the assessment of hydrologically related erosion processes in semiarid catchments of Southern Africa. In: Spatial methods for solution of environmental and hydrologic problems – science, policy, and standardization. ASTM International, West Conshohocken
Flügel WA, Märker M, Moretti S, Rodolfi G, Sidrochuk A (2003) Integrating geographical information systems, remote sensing, ground truthing and modelling approaches for regional erosion classification of semi-arid catchments in South Africa. Hydrol Process 17(5):929–942
Gao H, Birkel C, Hrachowitz M, Tetzlaff D, Soulsby C, Savenije HH (2019) A simple topography-driven and calibration-free runoff generation module. Hydrol Earth Syst Sci 23:787
Gober P (2018) Why is uncertainty a game changer for water policy and practice? In: Building resilience for uncertain water futures. Springer, Cham, pp 37–60
Graham LP, Andersson L, Horan M, Kunz R, Lumsden T, Schulze R, … Yang W (2011) Using multiple climate projections for assessing hydrological response to climate change in the Thukela River Basin, South Africa. Phys Chem Earth Parts A/B/C 36(14):727–735
Haigh E, Fox H, Davies-Coleman H (2010) Framework for local government to implement integrated water resource management linked to water service delivery. Water SA 36(4):475–486
Hawinkel P, Thiery W, Lhermitte S, Swinnen E, Verbist B, Van Orshoven J, Muys B (2016) Vegetation response to precipitation variability in East Africa controlled by biogeographical factors. J Geophys Res Biogeo 121(9):2422–2444
Herrfahrdt-Pähle E (2010) South African water governance between administrative and hydrological boundaries. Clim Dev 2(2):111–127
Hu Q, Su P, Yu D, Liu J (2014) Pattern-based wind speed prediction based on generalized principal component analysis. IEEE Trans Sustain Energy 5(3):866–874
Ikudayisi A, Adeyemo J (2016) Effects of different meteorological variables on reference evapotranspiration modeling: application of principal component analysis. World Acad Sci Eng Technol Int J Environ Chem Ecol Geol Geophys Eng 10(6):641–645
Jakob D (2013) Nonstationarity in extremes and engineering design. In: Extremes in a changing climate. Springer, Dordrecht, pp 363–417
Katz RW (2013) Statistical methods for nonstationary extremes. In: Extremes in a changing climate. Springer, Dordrecht, pp 15–37
Katz RW, Parlange MB, Naveau P (2002) Statistics of extremes in hydrology. Adv Water Resour 25 (8–12):1287–1304

Kim Y, Kim H, Lee G, Min K-H (2019) A modified hybrid gamma and generalized Pareto distribution for precipitation data. Asia-Pac J Atmos Sci 55(4):609–616

King JM, Tharme RE, De Villiers M (2000) Environmental flow assessments for rivers: manual for the building block methodology. Water Research Commission, Pretoria

Kundzewicz Z, Mata L, Arnell NW, Döll P, Jimenez B, Miller K, ... Shiklomanov I (2008) The implications of projected climate change for freshwater resources and their management. Hydrol Sci J 53:3

Liu X, Liu FM, Wang XX, Li XD, Fan YY, Cai SX, Ao TQ (2017) Combining rainfall data from rain gauges and TRMM in hydrological modelling of Laotian data-sparse basins. Appl Water Sci 7(3):1487–1496

MacFadyen S, Zambatis N, Van Teeffelen AJ, Hui C (2018) Long-term rainfall regression surfaces for the Kruger National Park, South Africa: a spatio-temporal review of patterns from 1981 to 2015. Int J Climatol 38(5):2506–2519

Makkeasorn A, Chang N-B, Zhou X (2008) Short-term streamflow forecasting with global climate change implications–a comparative study between genetic programming and neural network models. J Hydrol 352(3–4):336–354

Moges MM, Abay D, Engidayehu H (2018) Investigating reservoir sedimentation and its implications to watershed sediment yield: the case of two small dams in data-scarce upper Blue Nile Basin, Ethiopia. Lakes Reserv Res Manag 23(3):217–229

Najafi R, Kermani MRH (2017) Uncertainty modeling of statistical downscaling to assess climate change impacts on temperature and precipitation. Water Resour Manag 31(6):1843–1858

Null SE, Prudencio L (2016) Climate change effects on water allocations with season dependent water rights. Sci Total Environ 571:943–954

Null SE, Viers JH, Mount JF (2010) Hydrologic response and watershed sensitivity to climate warming in California's Sierra Nevada. PLoS One 5(4):e9932

Oyebode O, Adeyemo J, Otieno F (2014) Monthly stream flow prediction with limited hydro-climatic variables in the Upper Mkomazi River, South Africa using genetic programming. Fresenius Environ Bull 23(3):708–719

Paswan RP, Begum SA (2014) ANN for prediction of area and production of maize crop for Upper Brahmaputra Valley Zone of Assam. Paper presented at the advance computing conference (IACC), 2014 IEEE international

Poff NL (2018) Beyond the natural flow regime? Broadening the hydro-ecological foundation to meet environmental flows challenges in a non-stationary world. Freshw Biol 63(8):1011–1021

Schulze RE (1995) Hydrology and agrohydrology: a text to accompany the ACRU 3.00 agrohydrological modelling system. Water Research Commission, Pretoria

Schulze R, Pike A (2004a) Development and evaluation of an installed hydrological modelling system. Water Research Commission, Pretoria

Schulze R, Pike A (2004b) Development and evaluation of an installed hydrological modelling system. Water Research Commision report (1155/1), 04. Water Research Commission, Gezina

Shrestha PK (Producer) (2016, 14, May 2017) SWAT Calibration Helper v1.0 Microsoft Excel Worksheet. Retrieved from https://www.researchgate.net/publication/

Sohrabi MM, Benjankar R, Tonina D, Wenger SJ, Isaak DJ (2017) Estimation of daily stream water temperatures with a Bayesian regression approach. Hydrol Process 31(9):1719–1733

Sultan B, Janicot S (2000) Abrupt shift of the ITCZ over West Africa and intra-seasonal variability. Geophys Res Lett 27(20):3353–3356

Svensson C, Hannaford J, Prosdocimi I (2017) Statistical distributions for monthly aggregations of precipitation and streamflow in drought indicator applications. Water Resour Res 53(2): 999–1018

Tarasova L, Basso S, Zink M, Merz R (2018) Exploring controls on rainfall-runoff events: 1. Time series-based event separation and temporal dynamics of event runoff response in Germany. Water Resour Res 54(10):7711–7732

Taylor V, Schulze R, Jewitt G (2003) Application of the indicators of hydrological alteration method to the Mkomazi River, KwaZulu-Natal, South Africa. Afr J Aquat Sci 28(1):1–11

Taylor RG, Scanlon B, Döll P, Rodell M, Van Beek R, Wada Y, . . . Edmunds M (2013) Ground water and climate change. Nat Clim Chang 3(4):322–329

Tyson PD (1986) Climatic change and variability in southern Africa. Oxford University Press, Oxford

Ufoegbune GC, Yusuf HO, Eruola KO, Awomeso JA (2011) Estimation of water balance of Oyan Lake in the North West Region of Abeokuta, Nigeria. Br J Environ Climate Change 1(1):13–27. Retrieved from SCIENCEDOMAIN international www.sciencedomain.org

Viviroli D, Archer D, Buytaert W, Fowler H, Greenwood G, Hamlet A, . . . López-Moreno J (2010) Climate change and mountain water resources: overview and recommendations for research, management and politics. Hydrol Earth Syst Sci Discuss 7(3):2829

Wang S, Yu L, Tang L, Wang S (2011) A novel seasonal decomposition based least squares support vector regression ensemble learning approach for hydropower consumption forecasting in China. Energy 36(11):6542–6554

Wotling G, Bouvier C, Danloux J, Fritsch J-M (2000) Regionalization of extreme precipitation distribution using the principal components of the topographical environment. J Hydrol 233 (1–4):86–101

Yuan F, Berndtsson R, Uvo CB, Zhang L, Jiang P (2016) Summer precipitation prediction in the source region of the Yellow River using climate indices. Hydrol Res 47(4):847–856

4

Menace and Mitigation of Health and Environmental Hazards of Charcoal Production in Nigeria

Philip Olanrewaju Eniola

Contents

P. O. Eniola (✉)
Department of Agricultural Technology, The Oke-Ogun Polytechnic, Saki, Oyo State, Nigeria

Abstract

The use of biomass and biofuels, such as wood, charcoal, petroleum, kerosene, and gas, is becoming competitive based on the level of development of each nation. However, charcoal production (CP) and marketing now tends to be a major business among many households in both rural and urban communities with no consideration of its effects on climate change adaptation. While the research question considers the various definition of climate change adaptation, and the importance of charcoal production in Nigeria, the manuscript speaks mainly of the problems of charcoal production, the lack of planning to address these problems, and the lack of planning to move the communities away from this practice and out of poverty. It addresses the impacts of charcoal production on agriculture, such as lack or loss of labor and destruction of arable lands. The paper discusses the effect of charcoal production on health. Also, the environmental problems of CP are highlighted in the manuscript. The policy frameworks on forestry by the Federal Ministry of Environment 2006 with its flaws will be included. Remedy such as the establishment of a Land Use Planning Agency (LUPAG) and panacea for greening the charcoal value chain issues will be discussed. Lastly, attention is given to the agricultural adaptation strategies to climate change which are capable of reducing charcoal production, such as mixed cropping.

Keywords

Climate adaptation change · Poverty · Deforestation · Policy framework · Alternative to charcoal utilization

Introduction

Definitions of Climate Change Adaptation

Climate change adaptation refers to the process of modification of the present and future impacts of climate change (United Nations Climate Change 2020). Due to the rapidly changing climate, coupled with economic, social, and technological developments, nations have no option but to look for a way to achieve a greater stride which further creates more tension in the ozone layer. According to the Victorian Centre for Climate Change Adaptation Research (VCCCAR) (2020), several views are expressed by scholars as follows: adaptation to climate change is the initiative to enable people and their surroundings to withstand the changing climate conditions (United Nations Framework Convention on Climate Change (UNFCCC) (2015). The Intergovernmental Panel on Climate Change (IPCC) refers to climate change adaptation as tampering with the natural or human systems so as to respond to actual or expected climatic stimuli or their effects. However, the UN Development programs defines it as the process through which strategies to moderate, and withstand with the subsequent effects of climatic events that are developed, implemented and improved upon. Conversely, the UK Climate Impacts Program defines it as the outcome of a process that creates limitation to the distortion or risk of problem or actualization of benefits that are related to climate variability. But the National Climate change Adaptation Research and Facility (NCCARF) views climate change adaptation as a composition of actions undertaken to reduce the negative effects of climate change, exploiting any beneficial opportunities. Lastly, the Victorian Government states that it is a conscious step to prevent, manage, or decrease the effects of a warmer, drier, and more extreme climate and to maximize the benefits which such changes are likely to provide.

Importance of Charcoal Production in Nigeria

Nigeria is a country that is endowed with crude oil, natural gas, and energy resources, such as solar, wind, biomass, and biogas. The country is gifted with human resources with a total population of 140 million according to the 2006 population census. The annual population growth of Nigeria is about 2.8% (National Population Commission [NPC] 2006). Currently, the national energy supply is almost entirely dependent on fossil fuels and wood fuel (The World Forest Movement 2006). For instance, the value of the charcoal market for 26 sub-Saharan African countries is said to purportedly exceed $1.8 billion per year (Food and Agriculture Organization [FAO] 2018). In terms of energy, charcoal consumption in many African countries is higher than the quantity of electricity consumed. The importance of charcoal lies in the following areas.

It is a source of considerable amount of employment in rural areas (Arnold et al. 2006). A significant number of rural dwellers relies on charcoal production as a good source of employment, especially in a devastating economy like that of Nigeria. Money realized from charcoal represents per capita income of between 24 and 14 dollars in 1990 and 2000, respectively, which are equivalent to between 1.8 and 4.8 times of per capita income from the sale of agricultural produce in the same years. Charcoal producers reveal that the average annual production in some countries is 160 bags per producer, whereas some produce significantly more than 500 bags. Also, income from charcoal is expended on food, farm inputs, and implements (FAO 2018).

Charcoal production is also one of the recognized economic drivers in some African countries aside from agriculture. However, it generates more income for the rural dwellers (Williams 1993). It also allows for quick return on investments (Stefan 2009). About 65% of the people in the rural areas have made wood fuel and charcoal production a source of income (Shackleton et al. 2006) because of its quick return on investments, unlike the involvement in arable crop production that will bring income in 3 months. In addition to its export value, charcoal trade at the local level provides income opportunities to pay education, health, feeding, and ceremony bills for most rural and urban dwellers through small-scale retail and wholesale (FAO 2018, Fig. 1). It is a reliable, comfortable, and easy form of energy for heating and cooking, with a relatively cheaper cost (Iloeje 2002).

Fig. 1 Plate showing charcoal depot in Saki derived Savannah zone of Nigeria from field study on April 19, 2013

Charcoal is produced for the poor in rural and urban areas to enable them to meet their energy needs, such as for heating and cooking. It is the most recognized commercial fuel derived from wood. It is smoke-free and can be used in small quantities during cooking (Adam 2009). Charcoal can also produce greater heat than wood, and it is useful for many industries in urban environments. In most developing countries, it is the chief form in which wood fuel is used in towns NL Agency (2010). Charcoal can be stored, takes up little space than firewood for heat generation, and does not deteriorate so easily. It is easier to handle during transportation and distribution. Therefore, it is more preferable to wood (World Bank 2004).

Furthermore, charcoal is used in African art to design various objects (World Energy Council 2004). Charcoal is the raw material for the manufacture of gunpowder and industrial and automotive fuel and for purification/filtration of the cottage industries, the main reason why several individuals continue to produce charcoal. It is also used by blacksmiths and for other industrial applications (Chris 2007). Various uses of charcoal greatly influence the standard of living of rural people who do not have access to agricultural inputs and finances. Charcoal has a higher energy density compared with biomass fuels, and it can be stored without fear of being destroyed by insects (Seidel 2008).

Charcoal is a cheap source of energy for both rural and urban dwellers. However, modern economic activities depend mainly on petroleum products and electricity (FAO 2018). The difficulty associated with the production and use of charcoal, the growing urbanization, and the choice of urban dwellers is at a heavy drain on local wood resources (Arnold 2001).

Factors Responsible for the Production and Utilization of Charcoal

The factors that affect the production and utilization of charcoal in Nigeria.

Poverty are discussed as follows. The increase in the poverty level among Nigerians causes drives the production and use of charcoal. Although it is very essential for all the rural dwellers to use improved energy for cooking and heating, they lack the financial wherewithal to purchase kerosene, which is an alternative. They rely mainly on charcoal, which is cheap and available but has serious environmental hazards. Long gestation period of agricultural products in contrast to charcoal production that takes a lesser period of two or three weeks. Thus, rural dwellers prefer to use it as an alternative source of income to other agriculture income generating activities.

The lack of employment may encourage the mass movement of rural dwellers into charcoal production. The absence of job opportunities, especially paid ones, induces rural dwellers to participate in charcoal production as an alternative source of employment. The current economic recession in the country has made it possible to lay off both rank and file staffs in our various industries. Also, urban rehabilitation is forcing most motorcycle riders to relocate to the rural areas. This makes charcoal production an alternative source of employment.

The lack of awareness on the environmental effects of charcoal production is a serious issue. Most rural dwellers are not aware of the negative effects of charcoal production on their environment which, in turn, results in climate change. Poverty and level of education, among others, also influence awareness of the negative effects of charcoal production on the environment. Since the focus of every rural dweller is to obtain a means for survival, they are not concerned about the implications of their action.

Insufficiency of basic amenities in rural areas may trigger charcoal production. Rural dwellers usually depend on the government for their basic human needs, such as electricity, cooking gas, and the likes, in their domain. The inability of the government to provide such basic amenities triggers high production and consumption of charcoal as a perceived good source of energy for cooking and heating. Thus, the quantity of wood charcoal production in Nigeria stands at 4022763 tonnes per annum (World Bank 2011).

The lack of proper implementation and enforcement "by successive government" may affect our environment. There are good policies that have been established on forestry and allied matters in Nigeria. It is unfortunate that the change in government and loss of focus often inhibit their effective implementation. Sometimes, evaluation units to ensure strict implementation are absent. The focus of successive government always differs from one to another. However, irrespective of their different interests, every government is expected to focus on the mitigation process of climate change adaptation.

Industrial revolution: Before the discovery of oil in Nigeria, agriculture was the main occupation of about 70% of the Nigerian population, with most farmers having less than 1 ha of farming land. With the advent of industrialization, lands are cleared constantly. Thus, trees are cut down and used to produce charcoal.

The high cost of farming input can also increase the number of charcoal producers. Some rural dwellers lack the purchasing power for farm implements. As a result, they see charcoal production as a cheap means of fending for themselves when the cost of farming is high.

Effects of Charcoal Production on Agriculture

It should be noted that charcoal production has different effects on agriculture, such as lack or loss of labor. When the same labor that is expected to till the land for agricultural activities still cut down trees and makes charcoal kilns, farming activities are hindered, especially when they get quick income. It results in the reduction in the number of people in agriculture. When there are more enabling environments available for charcoal production, charcoal producers feel more comfortable in the business operation, thus decreasing the number of people involved in farming. This act will definitely lead to food shortage in the country. Charcoal production destroys arable lands. When trees are cut for charcoal production, the lands become susceptible to wind and water erosion. These negative practices eventually destroy the structure of the soil and make it infertile for arable crop farming.

Health Implications of Charcoal Production

Aside from the effects of charcoal production on the environment, there are also human health-related challenges (UNDP 2005). These include backache, heat, and cough, among other ailments, which are experienced a lot by charcoal producers. Repetitively moving heavy woods during charcoal production induces lumbar pain and muscular soreness to the producers (Tzanakis et al. 2001). It is worth noting that some charcoal producers lose their lives during the production. Accidents may also occur during the cutting of trees, kiln preparation, and loading of charcoal onto lorries. Furthermore, producers inhale gases and smoke, and the heat produced during the charcoal production is a source of ailments, such as respiratory diseases and cough. They also experience sore hands, fatigue, and chest pain. Sputum production, dyspnea, and hemoptysis are other ailments suffered by the producers.

Effects of Charcoal Production on the Environment

The major environmental problems caused by charcoal production in Nigeria are discussed as follows. Deforestation is a product of charcoal production which has significant impacts on the environment, especially with regard to increased erosion. It can also worsen climate change and threaten biodiversity. Deforestation is the destruction of forest areas for several purposes, such as agriculture, urbanization, and wood fuel and charcoal production. Charcoal production requires commercial felling of wood, thus leading to deforestation (Eniola 2014). Greenhouse gas (GHG) emission is a dangerous phenomenon of charcoal production. The reduction of forest cover also minimizes carbon consumption and results in the release of already-fixed carbon. For instance, the effect of charcoal on forest reserves is disastrous on two grounds. First, the wood fuel equivalent is four to six times greater due to the lack of a professional production process (Fig. 2) (SEI 2016). Second, emission of gas during charcoal production is significant compared with charcoal burning. Aside from the above, ecosystem destruction has a long-term negative effect on the environment. Woods are obtained illicitly from land, and charcoal producers are compelled to harvest woods to enable immediate charcoal production (Emeodilichi 2018). Charcoal producers may sometimes gather dead woods, twigs, and branches and allow trees to regenerate. Due to the cutting of live trees for charcoal production, some important wood fuel species, such as mahogany and shea tree, which were abundant in the past years, are becoming scarce (Pabi and Morgan 2002). Charcoal production also affects the soil structure in two ways: first, it impacts on the kiln site (soil) where combustion takes place as a result of huge heat that is released from the covering process, and second, it affects the surrounding/environment of the production pit. Ogundele et al. (2011) and Oguntunde et al. (2008) revealed that the soil in the site of charcoal production has a slight increase in pH. It reduces bulk density of soil surface temperature, higher infiltration rate, and surface albedo compare to places that are not subject to charcoal production (Chidumayoa and Gumbob 2012).

Fig. 2 Image of the combustion process during charcoal production in Saki from a field study conducted on April 19, 2013

Policy on Charcoal Production-Induced Environmental Problems in Nigeria

The major objective of the Federal Ministry of Environment on Forestry Policy (2006) is to achieve an acceptable level of self-sufficiency in wood products through the use of sound management methods. Hence, the main thrust of the policy is consolidation and increase in the forest estate and its management for the usage of the future generation. Due to the human-induced environmental problems, which the present forest estate cannot cope with, more forest estate is an area of land covered by different species of trees should be created in the country to prevent future disasters. Forest conservation and environmental protection are important. Our forests are no longer preserved adequately, unlike in the past. Cattle herdsmen can now be seen ravaging our forests, and wood fellers easily enter the forest and destroy the ecosystem. Thus, governments must exert more efforts to conserve our forests.

There is a great contrast between forest exploitation and forest regeneration. There should be a more increased rate of forest generation than forest exploitation to achieve a sustainable environment. There should be proper utilization of both the forests and its products. In addition, research on how non-forest products can take the place of forest products in all human activities, such as building and other energy requirements, need to be conducted. This will make the future generations benefit the forests and their products. Forest estates should be protected from fires, poachers, trespassers, and unauthorized grazers. It should also be noted that hunters, unauthorized grazers, and farmers endanger forest estates by setting them on fire to obtain their daily needs. These devastating habits must be prevented to secure the forests. The establishment of private forestry can significantly improve our forests. Thus, the government should permit and encourage private individuals to establish plantations of gmelina, teak, cashew, shear, and many other economic trees that can help to preserve and protect the enviroment. Also, the establishment of manmade forests for specific end-uses is a good means to regain the strength of forests. The government should establish a specific type of forest mainly for a particular tree user where they can go to without hindrance. The development of secondary forest products that are important in local economies and that support agro-forestry practices will boost the economy of the nation.

A policy that encourages the establishment of forestry to provide employment opportunities also needs to be implemented. The only thing that can reduce the occurrence of forest encroachment is to create more employment for people. Otherwise, more rural dwellers will rely on the forest and its products. In addition, more national parks and game reserves need to be created, and the existing ones should be properly maintained. An efficient use of wood energy and alternative sources of energy to wood fuel is also required. Lastly, cooperation among the international community with regard to forestry development must be encouraged (Federal Ministry of Environment 2006). Any law, regulations, and agenda in ensuring the sustainable use of the environment introduced by the international community must be strictly adhered to.

Problems with Policies and Practice of the 2006 Federal Ministry of Environment in Nigeria which Warranted for Its Improvement by the Federal Parliament in 2019

There are so many flaws in the 2006 forestry policies which the new one to be assented to by the President of Nigeria is expected to cater for. Among them is the alarming widespread and increasing land and vegetation degradation. The policy failed to prevent indiscriminate destruction of forests and their products. Forest users enter the forests freely and take away useful products and by-products. These nefarious acts result in fast land degradation. Every day, the influence of humans on the forest continues to increase due to the search for means of income. There are escalating real supply and demand deficits for forest products. The demands for forests and their products are greater than the available resources, which is a serious problem for the future generations. In Nigeria, credible databases are lacking, and there are no records of the quantity of woods and other products removed from the forest, of endangered species, and of future woods requirements. Moreover, the ineffective management of forest reserves is not given sufficient attention. The present system allows unauthorized grazers and poachers to enter the reserves freely. This may be due to the lack of adequate personnel, especially at the state level, and equipment to assist them in their duties. Negligible private investment in forestry is recorded in the implementation of the policy. Due to the several bottlenecks in acquiring land titles, private investors become discouraged in investing in forestry. There is abandonment of trees biodiversity that yield a low level of protection of existing reserves. Most of the forest users only prioritize their own needs at the expense of other forest products. And the protection of other forest products is being neglected. Moreover, there is a conflicting management of forests and its resources in the northern part of Nigeria. The states and local governments still compete for the control and utilization of the forest and its resources. This may be due to the fact that the environment is predominantly a grazing zone for cattle rearers.

The Climate and Development Knowledge Network (CDKN) of the UNFCCC (2015) expressed that it is very difficult for the less-developed countries, such as Nigeria, to implement any forestry policies that will give room to tangible climate adaptation strategies due to the following challenges: It is not easy to create awareness on the need and benefits of action among stakeholders, including policy makers. This may be linked to the high level of illiteracy in the country. A serious concern is how to enhance and integrate climate change into national planning and development processes. A budget allocation for climate change adaptation strategies every year can be beneficial. Also, the methods for strengthening the links between different tiers of government plans on climate change may not be easily realizable. This is because both segments of the government may develop different approaches to climate change adaptation without considering the global implications of such actions. Another method is to build capacity, analyze, develop and make use of climate change policy. There is a need for the inclusion of sound researchers in the implementation of climate change policy. How to establish a realistic mandate to coordinate actions around NDCs and drive their implementation is important to climate change adaptation in Nigeria. The best ways to address resource constraints are to develop and implement climate change policy. Both human and nonhuman resources may be considered as serious constraints, especially where expertise on climate and equipment are needed.

Factors that Make It Difficult to Plan for Alternative Environmentally Friendly Energy in Nigeria

Nigeria, in particular, and Africa, in general, are mostly affected by climate change due to indiscriminate felling of trees, poor management of agricultural lands, and other environmental pollution activities performed by man. It may be difficult to properly plan on the energy requirements of the country due to the lack of an updated population census of the country. It should be noted that the last census in the country was carried out in 2006, which reported an estimated 140 million population, which some regions in the country view as a fraud (NPC 2006). Yet, it was suggested that another census should have been held in 2016. Today, it is already 2020, and this year is not included in the budget allocation of the country. An average of about 200 million people is expected, but it is just an assumption that is yet to be scientifically proven. Thus, if energy (for cooking and heating) requirement of the country is to be adequately budgeted for, the population of the country must know otherwise charcoal production and utilization will continue to increase.

One serious factor is the lack of unique and strategic climate change adaptation policies. Apart from the 2006 forestry policies of the country which have not been adequately implemented owing to its deficiencies, no other specific climate change adaptation policies have been made. Since charcoal production constitutes serious environmental hazard to the climate of the nation, it is expected that climate change adaptation policies will cater for charcoal production and utilization must be put in place (FAO 2010). Included in the factors influencing the choice of alternative environmental friendly sources of energy in Nigeria is the lack of recent and accurate climate information that can be used in the planning for the required energy of the teeming population of the country. Even where it is available, it cannot be relied on because the modern facilities and equipment to be used are not available. The lack of an adequate subject matter specialist to interpret climate information is also an obstacle. The government or relevant agencies cannot afford the required modern climate facilities; thus, effective climate change adaptation strategies cannot be developed in the country. This will also affect the effective dissemination of such data that can be used by farmers, charcoal producers, and other land users for future disasters. Another obstacle is the poor investment in scientific research on the impacts of climate change. It is the amount of human capital invested in the study of climate change impact by relevant professions to enhance the knowledge on how to address the challenges of climate change in the country. Another factor is that most African countries have weak governance. It should be noted that all the three tiers of our government are very weak in terms of the implementation of climate change adaptation policies. Since the federal government is weak, the local government abuses the forest resources. There is no effective monitoring at the state and local levels on climate change adaptation policies, which further creates serious damage to the climate.

Strategies to Review the Existing Forest Policy on the Provision of an Enabling Environment for Sustainable Forestry Management and Development in Nigeria

To achieve an effective forestry management in Nigeria, the following strategies must be implemented. A land-use planning agency (LUPAG) should be established, and it should be noted that the existing land-use plan attached to the Ministry of Lands is obsolete and cannot cope with the trend of development in the country; hence, it should be an autonomous agency. This will enable the agency to really find immediate and long-term solutions to the problems on land utilization in the country.

The forest sector policy should be redefined considering the concepts of cognizance modern forestry. In the past, individuals did not own a forest. People now acquire lands for the purpose of establishing plantation. Land tenure systems and tree onwership should be reviewed. This will make it easier for everyone to have unfettered access to land and improve forestry development in Nigeria. In the ancient tradition of land ownership, a female does not receive support to own land even by inheritance. This practice limits their ability to use their resources to acquire wealth. Emerging products and services should also be considered. Before, only few airports, railways, and other transport services were available in the country. With the increase in population, the demand for these services will also increase, which will definitely have a serious effect on forest resources (FAO 2011). The foresrty policy should embrace corporate protection and management measures that will give both the states and the local government areas absolute control of the forest resources. It has been noted that the local governments mostly affected by deforestation do not even bother to campaign for activities to mitigate this. Hence, specific roles should be given to states and local governments.

Another adaptation strategy includes the implementation of law on the regulation of felling of trees. This law exists but is yet to be fully felt. Fellers have a free day in the forest to the extent that they cut economic trees, such as shea, locust bean, mango, and cashew trees, among others. Thus, the "cut-one, plant four" approach should be adopted. All tree fellers must be required to plant four trees as substitutes for a tree that is cut down on their personal lands. Aside the use of governement agencies as task force, various community heads should be involved in tree head count and serve as a force to implement the approach. Regulations for the exportation of charcoal are imperative. It should be borne in mind that the consumption of charcoal in Nigeria is lesser than the annual exportation. Europe is taking much of Nigeria's charcoal, and this has been made possible by the high amount of money paid to the suppliers. In addition, seaports and dry ports in Nigeria were also not properly monitored (Mbura 2015). Monitoring, inspection and enforcement of the modern method of charcoal production processes: charcoal producers still use obsolete means to produce charcoal and it has adverse effect on the environment in that it causes pollution and endangers the ozone layer (Bailis 2009; Bailis et al. 2015). In charcoal production, sustainably managed resources, such as natural forests, planted forests, and community forests, as well as improved technologies can be used to reduce GHG emissions (Adger et al. 2003; Olson 2009; Vos and Vis 2010). The use of other alternatives to energy, such as agricultural waste and residues of wood and twig outside forests, as well as agroforestry will be advantageous. It is noteworthy that wastes from various crops, such as maize, guinea corn, wheat and oat husks, chaffs of rice, and cassava peels, are left unused and are thus burnt instead of being converted to usable energy. Charcoal dust should be converted into briquettes. Large tonnage of the dust which can be turned into charcoal are left on charcoal production sites and loading stores. However, the rural charcoal producers lack the skills to do this. Better management of local kilns increases efficiency of charcoal production. During the monitoring stage, charcoal producers may not regularly examine the extent of the smoke formed, thereby giving the kiln the opportunity to burn without complete combustion (Otu-Danquah 2010). The use of fossil fuel for transportation should be minimized. The country should research on other energy sources aside from fossil fuel for both household and industrial energy requirements. Cooking energy such as gas should be optimally utilized. If this is achieved, the cost of cooking gas should be within the reach of the poor in rural areas.

Climate Change–Charcoal Chain Sector: The Policy Options

Charcoal climate adaptation value chain is in dire need of policy implementation.

Various government agency regulatory bodies environments are used by charcoal value chain and should be harnessed effectively (FAO 2017). Based on this, policies must be formulated to improve charcoal production technology that will further enhance the sustainable deployment of resources in the forest. Moreover, there is need for an enduring policy projection that will enhance the sustenance of charcoal value chain through diversification and democratization of clean energy alternatives to lessen the increasing demand for charcoal. Incentives, reward distribution, meaningful management of forestry resources, land use planning, landscape management, and a green economy are important to greening the charcoal value chain (Minten et al. 2013). Technological advancement is imperative due to the differences in taxation that could have assisted the incentivization of sustainable sourcing and production of charcoal through fees and licenses. Global fiscal devices related to climate change reduction, such as the clean development mechanism, and reducing emissions from deforestation and forest degradation can provide financial boosts. Effective forest law enforcement and administration are important to boosting government income generation and investments through sustainable forest management and efficient techniques for wood conversion (Sander et al. 2013). To limit the excess of charcoal producers, in May 2016, Nigeria banned charcoal production when producers and exporters failed to adhere to the cut-one, plant four policy. It was not more than 2 months when a serious outcry from the populace crippled the restriction since charcoal is the major source of energy in the country. "The problem is complicated because the Federal Government owns the policy and the machinery to enforce the law, but the states own the forest. The states also see logging as a form of revenue generation." Since the government fails in that regard, a serious alternative, such as solar energy, has been targeted as it is environmentally friendly and economically viable. The big question is, how many people could afford it? Nigeria's National Council on Environment reiterates that the country needs an effective policy to promote solar energy and efficient cooking stove. Although the focus on renewable energy is a right step forward, its success will largely depend on how Nigeria is able to sensitize the people to adopt it. Improved monitoring of rural dwellers can enhance their willingness and preparation to invest in enduring methods. Traditional leaders still hold much title on land acquisition even more than the government. They place difficult conditions on land acquisition which prevent long-term investment in forestry. The transfer of responsibility and provision of financial and human resources to grassroot authorities can boost sustainable forest management and charcoal production. Formulation of policies encourages private sector participation in the dissemination of improved technologies as well as the establishment of marketing system for sustainable products. The government, private sectors, producers, and consumers have a lot to gain through proper planning and decision-making process for charcoal management. Openness in revenue channels and accountability of all players in charcoal production are important to the growth of national and local economies. A solid institutional framework comprising of forest managers, tree growers, charcoal processors, and traders is inevitable to achieve effective coordination of initiatives that will help develop a sustainable charcoal value chain as well as elucidate the mandates of the stakeholders. Charcoal chain reform should establish a firm relationship among major actors. It should be conscious of the danger of corruption as well as protection of the few policies that are designed for the regulation and improvement of the value chain. Measures that will help secure and protect the rights to energy access of those who do not have options should also be considered. For instance, despite the fact that the 2013 European Timber Regulation (EUTR) stipulates that there should be no illegal exportation of any wood products to its members countries, they do not monitor charcoal importation. Charcoal bags are labeled with tags, such as sustainably cultivated, regardless of the type of wood the charcoal is made from.

Panacea for Greening Charcoal Value Chain

To improve the charcoal value chain, multiple concurrent interventions should be promoted to significantly lessen GHG emissions (Beukering et al. 2007). A green charcoal value chain needs financial viability that is best guaranteed through the improvement of tenure arrangements and legal access to resources. This will improve the purchase of wood and other biomass for charcoal production. It will also ensure the merits of a green charcoal value chain for the economies in the countries by placing financial premium on wood resources and incentivizing long-term practices and attractive investments that will induce a smooth movement. A comprehensive national foresrty policy framework should be developed for its sustainable administration and integration of charcoal with greater efforts across sectors to reduce the incidence of climate change. This will make climate change specific component of NDCs. Furthermore, governments and other key players should be supported in greening the charcoal value chain through research contribution. There should be holistic methodical examinations of charcoal value chain in major countries where charcoal is produced. In addition, information on GHG emissions at diverse levels of the charcoal value chain should be available (Bailis et al. 2005) as it will facilitate in the greening of the charcoal value chain. The role of charcoal production in deforestation and forest degradation, including other forms of deforestation and forest degradation drivers in urban should be assesed at everytime before leading to devastated level; socio-economic and environmental results and trade-offs of a green charcoal value chain at the local, subnational, national, and regional levels as well as the spread, results from pilot projects, success stories, and research within the entire charcoal value chain are essential.

Agricultural Adaptation Strategies to Climate Change that Can Reduce Charcoal Production

Great efforts must be exerted to encourage rural dwellers to go back to farming amidst the climate change that has destroyed the natural environment. Various agricultural adaptation strategies must be adopted to manage the soil resources such as mixed cropping. Mixed cropping is the practice of growing two or more crops together in the same field. It was derived from the traditional method of utilizing land particularly where there is shortage of land. For instance, cereals such as maize, sorghum, and legumes (cowpea and groundnuts) can be planted together. Mixed cropping is advantageous in that it varies in the period of maturity such as maize and cowpea, drought tolerance (maize and sorghum), input requirements (cereals and legumes) and final users of the products (e.g., maize as food and castor oil for cash). Hence, farmers should be encouraged to practice mixed cropping to maximize lands. Improved irrigation efficiency is required to make farming a business throughout all the seasons. Achievement in climate change adaptation depends on the access to adequate water in drought-prone areas. Knowing full well that water can be an inhibiting factor, improved irrigation efficiency is a significant adaptation device toward realizing food demands, especially during off seasons. Harvest during the dry season usually attracts good market price compared with that during the rainy season (Orindi and Eriksen 2005). Thus, there is a need for the adoption of soil conservation measures. Soil conservation deals with the proper timing of various farming activities that demand local experiences, burying of crop residues to improve soil fertility, burning of crop wastes to achieve quick release of nutrients, and allowing animals to graze on farmlands after harvesting crops to improve soil organic matter. Mulching is also important for controlling soil temperatures and extreme water loses. In addition, it prevents the emergence of diseases and harmful pests and conserves soil moisture. Before the advent of chemical fertilizers, rural farmers heavily depended on organic farming, which reduces GHG emissions. Planting of trees (afforestation) and agroforestry are mechanisms of soil preservation. Tree planting is the technique of transplanting seedlings for the following purposes: forestry, land reclamation, and landscaping. Tree planting in silviculture is called reforestation or afforestation, depending on whether the area that is used for planting is or is not recently forested. It involves planting of seedlings over an area of land where trees and plants have been removed by man, fire, or pest and diseases. The free distribution of tree seedlings to farmers to achieve quick afforestation is a method of adapting to climate change (Akinnagbe and Irohibe 2014; Bird et al. 2011). With the reduction in the quantity of available water for crop production, there is a need for crop breeders to provide varieties of crops that are resistant to drought. These crops can be planted in drought-prone areas to lessen their vulnerability to climate change. For example, oat and wheat require fairly less irrigation water compared with dry-season rice. Furthermore, drought resistant crop varieties have been subjected to test by smallholder growers as adaptation methods to climate change in Nigeria, Senegal, Burkina Faso, and Ghana (Ngigi 2009). Furthermore, crop diversification should be encouraged. medium or long time crop production is a possibility in diversification to high-value crops. The diversification of crop planting is an adaptation mechanism prioritized in areas where irrigation is used or not used. Thus, farmers should be encouraged to change the type of crop they plant to curb the risk of low yield from harvest on farm (Orindi and Eriksen 2005; Adger et al. 2003). Crop diversification can be a buffer against variations in rainfall. In addition to the changes in cropping patterns and planting calendars, long-term alterations in rainfall due to climate change can negatively impact crop production. Thus, it is important for extension agents to orient rural farmers. Such an orientation will enable the farmers to familiarize the trend of cropping patterns in their domains. Essentially, shifting from charcoal production to other livelihood activities will prevent our environment from further deterioration. Most rural dwellers consider charcoal production as their last resort to generate income (Eniola 2014). Hence, there is a need to create awareness on other livelihood activities, such beekeeping, mushroom farming, fish farming, and horticultural farming, among others, with special financial support from the government and other non-governmental agencies.

Conclusion and Recommendations

This chapter concludes that the government in Nigeria needs to exert more effort on climate change adaptation, especially in terms of the eradication or reduction of the threats posed by charcoal production to human health and the environment. Charcoal production cause severe damage to both the environment and humans, thus leading to the reduced number of people engaging in farming activities. Quick actions in ensuring impeding food shortage and rebuilding the forest must be taken so as not to jeopardize the future need of the forests by the coming generation. The country should strictly adhere to the laws, regulations, and policies of the international bodies on climate change adaptation. Alternative means of livelihood, such as fish farming, mushroom production, vegetable production, and other simple and cheap farming activities, should be embraced by farmers. The reduction of the number of rural dwellers involved in charcoal production will go a long way.

References

Adam JC (2009) Improved and more environmentally friendly charcoal production system using a low-cost retort- kiln (eco- charcoal). Renew Energy 34:1923–1925. Google Scholar

Adger WN, Huq S, Brown K, Conway D, Hulme M (2003) Adaptation to climate change in the developing world. Progress Dev Stud 3:179–195. Google Scholar

Akinnagbe OM, Irohibe IJ (2014) Agricultural adaptation strategies to climate change impacts in Africa: a review. Bangladesh J Agric Res 39(3):407–418. Google Scholar

Arnold MJE (2001) Forestry, poverty and aid. CIFOR occasional paper no. 33. Centre for International Forestry Research, Bogor. Google Scholar

Arnold JEM, Köhlin G, Persson R (2006) Woodfuels, livelihoods and policy interventions: changing perspectives. World Dev 34(3):596–611. Google Scholar

Bailis R (2009) Modeling climate change mitigation from alternative methods of charcoal production in Kenya. Biomass Bioenergy 33:1491–1502. Google Scholar

Bailis R, Ezzati M, Kammen DM (2005) Mortality and greenhouse gas impacts of biomass and petroleum energy futures in Africa. Science 308:98–103. Google Scholar

Bailis R, Drigo R, Ghilardi A, Masera O (2015) The carbon footprint of traditional wood fuels. Nat Clim Chang 5:266–272. https://doi.org/10.1038/nclimate2491. Google Scholar

Beukering PJH, Kahyararab G, Masseya E, di Primaa S, Hessa S, Makundi V, van der Leeuw K (2007) Optimization of the charcoal chain in Tanzania. PREM (Poverty Reduction and Environmental Management working paper) 07/03. Institute for Environmental Studies, Amsterdam. Google Scholar

Bird ND, Zanchi G, Pena N, Havlík P, Frieden D (2011) Analysis of the potential of sustainable forest-based bioenergy for climate change mitigation. Working paper no. 59. Bogor Indonesia Center for International Forestry Research (CIFOR), Bogor. Google Scholar

Chidumayoa EN, Gumbob DJ (2012) The environmental impacts of charcoal production in tropical ecosystems of the world: a synthesis. Energy Sustain Dev 17:86. https://doi.org/10.1016/j.esd.2012.07.004. Google Scholar

Chris P (2007) The age of wood: fuel and fighting in France forests. 1940–1944, Retrieved September 10, 2009, from http://www.histroycooperative.org/journals/ch/11.4/pearson.html

Emeodilichi HM (2018) Assessment of charcoal production processes and the environment impact in Kaduna, Nigeria. Resour Environ 8(5):223–231. https://doi.org/10.5923/j.re.20180805.02. Google Scholar

Eniola PO (2014) Perceived environmental and health effects of charcoal production among rural dwellers in agro-ecological zones of Nigeria. U.I PhD thesis. 2014. Google Scholar

FAO (2010) What woodfuels can do to mitigate climate change. FAO forestry paper no. 162. FAO, Rome. Google Scholar

FAO (2011) Framework for assessing and monitoring forest governance. FAO and the Program on Forests (PROFOR), Rome. Google Scholar

FAO (2017) The charcoal transition: greening the charcoal value chain to mitigate climate change and improve local livelihoods by J. van Dam. Food and Agriculture Organization of the United Nations, Rome. Google Scholar

FAO (2018) FAOSTAT ON CHARCOAL, Food and Agricultural Organization of the United Nations, 2017, pp 34–45. Google Scholar

Federal Ministry of Environment (2006) Forestry policies, pp 24–78. Retrieved 12th January 2021from http://www.fao.org/forestry/15148-0c4acebeb8e7e45af360ec63fcc4c1678.pdf. Google Scholar

Iloeje OC (2002) Renewable energy development in Nigeria: status and prospects. Ed. O.E. Ewart. Proceedings of a National Workshop on Energising Rural Transformation in Nigeria: Scaling Sup electricity Access and Renewable Energy market Development. Federal Ministry of Power and Steel, Abuja Nigeria. March 19–20 2001, 180. ICEED

Mburia R (2015) Africa climate change policy: an adaptation and development challenge in a dangerous world. Climate Emergency Institute, UNEP, Kenya Google Scholar

Minten B, Sander K, Stifel D (2013) Forest management and economic rents: evidence from the charcoal trade in Madagascar. Energy Sustain Dev 17(2):106–115. Google Scholar

National Population Commission (2006) Census news: a house magazine of the 2006 National Population Commission, pp 1–88. Retrieved 13th January, 2021 from https://nigeria.opendataforafrica.org/ifpbxbd/state-population-2006. Google Scholar

Ngigi SN (2009) Climate change adaptation strategies: Water resources management options for smallholder farming systems in Sub-Saharan Africa. The MDG Centre for East and Southern Africa, the Earth Institute at Columbia University, New York, p 189. Google Scholar

NL Agency (2010) Making charcoal production in Sub Sahara Africa sustainable. NL Agency and BTG Biomass Technology Group BV, Utrecht. Google Scholar

Ogundele AT, Eludoyin OS, Oladapo OS (2011) Assessment of impacts of charcoal production on soil properties in the derived savanna, Oyo state, Nigeria. J Soil Sci Environ Manage 2(5):142–146. ISSN 2141-2391 ©2011 Academic Journals. Google Scholar

Oguntunde PG, Abiodun BJ, Ajayi AE, Giesen N (2008) Effects of charcoal production on soil physical properties in Ghana. J Plant Nutr Soil Sci 171:591–596. Google Scholar

Olson AR (2009) A smoke burner for charcoal kilns. Northeastern longer 1941. A portable charcoal kiln. Connecticut Agric. Experiment Station Bull 448 New Haven. Google Scholar

Orindi VA, Eriksen S (2005) Mainstreaming adaptation to climate change in the development process in Uganda. Ecopolicy series 15. African Centre for Technology Studies, Nairobi. Google Scholar

Otu-Danquah KA (2010) Current status of charcoal demand and supply, and initiatives on improved cook-stoves. A presentation made during a kickoff meeting for TEC/ESMAP survey on the energy access and productive uses for the urban poor, held in the SSNIT Guest House Conference Room, Accra. On 11 Aug 2010, p 23. Google Scholar

Pabi O, Morgan EA (2002) Land-cover change in the Northern Forest-Savannah Transition in Ghana, commissioned technical report for the NRSP R7957 project. Retrieved April 23 from www.nrsp.org/pubs/index.rsp. Google Scholar

Sander K, Gros C, Peter C (2013) Enabling reforms: analyzing the political economy of the charcoal sector in Tanzania. Energy Sustain Dev 17(2):116–126. Google Scholar

Seidel A (2008) Charcoal in Africa importance, problems and possible solution strategies. Deutsche Gesellschaft für Technische Zusammenarbeit (GTZ) GmbH, Household Energy Programme – HERA Eschborn, Apr 2008, pp 28–35. Google Scholar

Shackleton CM, Shackleton SE, Buiten E, Bird N (2006) The importance of dry woodlands and rainforests in rural livelihoods and poverty alleviation in Southern Africa. Rainforest Polit Econ 9(2006):558–577. Google Scholar

Stefan C (2009) Fundamentals of charcoal production. National bioenergy center IBI conference on biochar, sustainability and security in a changing climate, pp 68–92. Google Scholar

Stockholm Environment Institute (SEI) (2016) How Kenya can transform the charcoal sector and create new opportunities for low-carbon rural development. Discussion brief. Stockholm Environment Institute. (SEI). Available at www.sei-international.org/mediamanager/documents/Publications/SEI-UNDP-DB-2016-Kenya-sustainable-charcoal.pdf. Google Scholar

The World Forest Movement (2006) State of charcoal in Nigeria economy: Effects on governance. Upland Press, p 8

Tzanakis N, Kallergis K, Bouros EB, Samiou FS, Siafakas NM (2001) Short-term effects of wood smoke exposure on the respiratory system among charcoal production workers. Chest 119(4):1260–1265. Google Scholar

United Nations Climate Change (2020) 2020 United Nations Framework Convention on Climate Change. Retrieved from http/www.unfccc.int/topics/adaptation-and-resilience/the.... Google Scholar

United Nations development Programme (UNDP) (2005) Basing national development on the millennium development goals. Retrieved August 20, 2008, from http://www.cifor.cgiar.org. Google Scholar

United Nations FCCC (2015) United Nations framework convention on climate change. Retrieved September 28, 2020 from http://unfccc.int/resource/docs/2015/cop21/eng/l09r01.pdf. Google Scholar

VCCCAR (2020) Victorian Centre for Climate Change Adaptation Research. Climate change adaptation definitions. Retrieved June 11 2020 from http://www.vcccar.org.au/climate-change-adaptation-definitions. Google Scholar

Vos J, Vis M (2010) Making charcoal production in Sub Sahara Africa sustainable. NL Agency, Utrecht. Google Scholar

Williams A (1993) An overview of the use of woodfuels in Mozambique and some recommendations for biomass energy strategy. National directorate of forestry and wildlife/biomass energy unit, p 250

World Bank (2004) Harvesting opportunities. Rural development in the 21st century. IV Regional Thematic Forum. Printed from the World Bank Group. Latin America and The Caribbean, p 248. Google Scholar

World Bank (2011) Economics of adaptation to climate change. Retrieved June 11 2020 from http://www.worldbank.org/en/news/feature/2011/06/06/.... Google Scholar

World Energy Council (WEC) (2004) Comparison of energy systems using life cycle assessment. London, p 45. Google Scholar

5

Barriers to Climate Change Adaptation Among Pastoralists: Rwenzori Region, Western Uganda

Michael Robert Nkuba, Raban Chanda, Gagoitseope Mmopelwa, Akintayo Adedoyin, Margaret Najjingo Mangheni, David Lesolle and Edward Kato

Contents

M. R. Nkuba (✉) · R. Chanda · G. Mmopelwa · D. Lesolle
Department of Environmental sciences, Faculty of Science University of Botswana, Gaborone, Botswana
e-mail: mnkuba@gmail.com; chandar@mopipi.ub.bw; gmmopelwa@mopipi.ub.bw; david.lesolle@mopipi.ub.bw

A. Adedoyin
Department of Physics, Faculty of Science, University of Botswana, Gaborone, Botswana
e-mail: akintayo_adedoyin@yahoo.com

M. N. Mangheni
Department of Extension and Innovation Studies, College of Agricultural and Environmental Sciences, Makerere University Kampala, Kampala, Uganda
e-mail: mnmangheni@gmail.com

E. Kato
International Food Policy and Research Institute, Washington, DC, USA
e-mail: E.Kato@cgiar.org

Abstract

This chapter discusses the barriers to climate change adaptation among pastoralists in the Rwenzori region in Western Uganda. Despite the implementation of adaptation programs by public and private agencies, pastoralists still have impediments to adapting to climate change. Data was collected using a household survey involving 269 pastoralists. The results revealed that the main barriers were poor access to climate change information, poor access to extension services, high cost of adaptation measures, poor access to credit, and insecure land tenure. There is need to improve capacity building of extension workers and other stakeholders in the dissemination of climate change information. Land tenure and land rights issues should be given high consideration in climate change adaptation policies and programs. Climate finance programs should be made more effective in addressing the high cost of adaptation.

Keywords

Barriers · Adaptation · Climate change · Pastoralists · Rwenzori · Uganda

Introduction

Climate change is one of the greatest threats to achieving the sustainable development goals related to eradicating poverty and hunger and ensuring clean water and life on land in Africa (IPCC 2018c). Despite the increased efforts at both national and international levels in the form of climate finance and programs, pastoralists still have barriers to adapting to climate change. Failure to adapt to climate change has resulted into high livestock mortality leading to a poor quality of life among pastoral households (IPCC 2014b). The impacts of global warming of 1.5 °C above pre-industrial levels are likely to lead to an increase in droughts and floods in Africa, especially in pastoral areas (IPCC 2018a). This could result in a failure to achieve sustainable development goals (IPCC 2018c). Pastoralists have adapted to climate change with various strategies that include livestock diversification, destocking, livestock migration, and engagement in nonfarm enterprises (Greenough 2018; IPCC 2014b). However, some pastoral households have been more vulnerable than others to climate change, resulting in loss of livestock (Greenough 2018). This implies that there are hindrances to adaptation.

Developing countries such as those in Africa still have challenges in addressing the hindrances, leading to an adaptation deficit when compared to developed countries in Europe (IPCC 2014a; Shackleton et al. 2015). These hindrances are categorized as limits and barriers. Barriers refer to impediments that can be easily overcome, while limits refer to impediments that cannot be easily overcome (IPCC 2014a). Many scholars have written about the barriers to adaptation without scrutinizing their causes (Filho and Nalau 2018; Shackleton et al. 2015). Moser and Ekstrom (2010) argue that barriers have various causes which have to be addressed for successful adaptation to be achieved. The issues of concern include the context of the pastoral household, type of pastoral household, and the governance system of the natural resource. This suggests that smallholder pastoralists experience different barriers compared to large-scale pastoralists. Rangeland policies that promote sedentary pastoralism have different barriers compared to those that promote mobile pastoralism (Löf 2013; Shackleton et al. 2015). These policies have strong implications for the governance system and context of pastoralism. For example, sedentary policies in pastoralism have a detrimental effect on the land tenure system that does not promote herd mobility as a coping mechanism for droughts (Little et al. 2008). This then becomes a barrier.

The barriers that have been identified include economic, biophysical, financial, informational, sociocultural, governmental, and institutional (Alam et al. 2018; Antwi-Agyei et al. 2015; IPCC 2014a; Shackleton et al. 2015). Juana et al. (2016) reported that barriers identified among livestock farmers in Botswana included poor access to improved technology; lack of land, extension workers, credit, and markets; and government restrictions on land use. Muller and Shackleton (2014) showed that the main barriers to commercial livestock farmers in semiarid East Cape in South Africa were poor access to finance, lack information on climate change adaptation and climate information, and lack of government support. This shows that there are barriers that are unique to livestock farmers and pastoralists.

The "barrier to adaptations" diagnostic framework was used to analyze the causes of the barriers (Moser and Ekstrom 2010). This was in order to have a robust investigation of the causes of the barriers, leading to measures of overcoming them. The analysis took into account the sources of the barriers that include the pastoralists (actors), context (governance and economic setting), and socio-ecological system in the rangelands (system of concern) (Moser and Ekstrom 2010). The larger context refers to the level of socioeconomic development and the control of information flow at national and local government level associated with climate change adaptation. Governance refers to the laws, policies, implementation frameworks, and resource allocation at national and local government levels that are associated with the socio-ecological system. The socio-ecological system refers to rangelands, herdsmen, and pastoral households. The socio-ecological system produces signals of environmental change (Moser and Ekstrom 2010) in terms of floods and droughts, although droughts have slow onset and may not be easily detected by the pastoralists. Early warning systems such as seasonal climate forecasts that are disseminated via mass media are not usually observed (Luseno et al. 2003). Pastoralists tend to use indigenous forecasts in their adaptation to extreme weather events despite their reliability and accuracy being affected by climate change (Speranza et al. 2010). Some pastoralists use both indigenous and scientific forecasts (Lybbert et al. 2007). The analysis also took into consideration the temporal dimension of the barrier sources, which include recent occurrence (contemporary issue) or occurrence over a long time (legacy issue) (Moser and Ekstrom 2012). The purpose was to find out when the barriers came into play. The spatial jurisdiction origin of the barrier in relation to pastoralists includes proximate, referring to the origin of the barrier being within pastoralists' cycle of influence, and remote, referring to the origin of the barrier being outside the pastoralists' cycle of influence, for example, government policies (Moser and Ekstrom 2010). The idea was to examine which barriers were within the cycle of influence of the pastoralists.

This chapter examines the barriers to adaptation among pastoralists in the Rwenzori region in Western Uganda. The key question is what are barriers to adaptation among pastoralists? The sources and origins of the barriers are examined to provide strategies for overcoming them. The objective is to generate empirical evidence that could be used by policymakers, development partners, extension workers, and nongovernmental organizations to address the barriers among pastoralists to improve their adaptation to climate related risks. The chapter contributes to pastoral literature that relates to barriers to climate change adaptation in Africa. The scope of this chapter addresses barriers among pastoralists in the tropical equatorial region in Western Uganda and does not cover arable farmers and pastoralists in semiarid areas.

Methods

Study Area

The study on barriers to adaptation among pastoralists was conducted in Kasese and Ntoroko districts found in the Rwenzori region in Western Uganda (Fig. 1). There are a few weather stations used to provide the climate data of the region. The rangelands in the study area are conducive to pastoralism and wildlife conservation, and wildlife protected areas (WPAs) exist in the area. WPAs include Queen Elisabeth National Park in Kasese district and Tooro-Semiliki Game Reserve in Ntoroko district (Fig. 1). As an adaptation strategy for climate change, pastoralists migrate to the eastern part of the Democratic Republic of Congo. Political instability in Eastern DR Congo tends to cause large numbers of pastoralists to migrate back to Uganda (KRC and RFPJ 2012). However, some pastoralists illegally graze in WPAs. The Uganda Wildlife Authority imposes heavy penalties on pastoralists who illegally graze in WPAs. The emphasis on sedentary pastoralism in government rangeland policies has led to a reduction in mobile pastoralism in Uganda (Wurzinger et al. 2006, 2009). The region has a bimodal rainfall distribution and experiences droughts and floods with increased frequency (NAPA 2007).

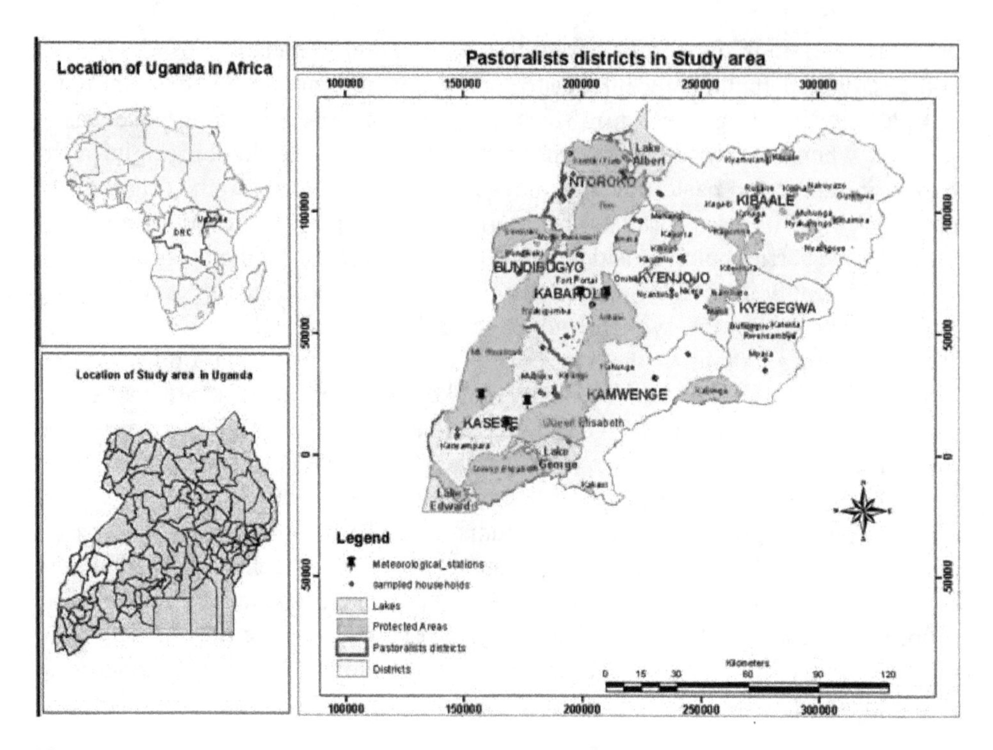

Fig. 1 Location map of study area. (Source: Michael Robert Nkuba)

Data Collection Methods and Sample Size

Data collection took place from August to October 2015 in the Rwenzori region. Household surveys gathered data on barriers to adaptations, socioeconomic characteristics, and use of indigenous forecasts (IFs) and/or scientific forecasts (SFs). A two-stage stratified sampling design was used (Cochran 1963), in which the strata were the districts and the second stage units were households. Stratified sampling was based on farming systems and agro-ecosystems in the Rwenzori region. Random sampling was used to select the respondents for the survey. The sample size was 778 households, with a 95% confidence level and a margin of error of 3.5%, based on the total population of the study area (the Rwenzori region and the Kibale district) of 102,496 households, according to Uganda population census report 2014. However, to allow for replacement in the sample of those who might back out of the study, and have good sizes for subsamples for those who use IF and SF and IF only, 19% of the statistically selected sample was included, giving a total study sample of 924. This was also to ensure a good sample size for subsamples (for those who use IF and SF and IF only). After data cleaning, 17 questionnaires were excluded from the analysis due to incomplete responses. Of the remaining 907 respondents, 580 were arable farmers, 269 pastoralists, and 57 agro-pastoralists. For this chapter, the sample size is 269 from pastoral areas in Kasese and Ntoroko districts. Data was analyzed using Stata 12. The scope of this chapter addresses barriers among pastoralists in the tropical equatorial region in Western Uganda and does not cover arable farmers and pastoralists in semiarid areas.

Results and Discussion

Socioeconomic Characteristics

The descriptive characteristics show that respondents were mostly male (59%) (Table 1) with an average number of local cows produced in the previous last 12 months being 70 per pastoral household. Pastoralists mainly used both IFs and SFs (59%) or IFs only (41%) in their adaptation to climate-related risks (Table 2). The use of indigenous forecasts has enhanced climate change adaptation and disaster

Table 1 Socioeconomic characteristics of respondents

Variable	Variable definition	Full sample ($N = 269$)
Female	Gender of the respondent (1 if female)	0.41
Male	Gender of the respondent (1 if male)	0.59
No school	Respondent had no formal school education (1 if yes)	0.38(0.50)
Primary	Respondent attained primary education (1 if yes)	0.47(0.50)
Secondary education	Respondent attained ordinary or advanced secondary education (1 if yes)	0.12(0.33)
Farm experience	Farming experience of the respondent in completed years	27.72(13.84)
Age	Age of the respondent in completed years	44.26(13.25)
Kasese	Respondent resides in Kasese district (1 if yes)	0.17(0.38)
Ntoroko	Respondent resides in Ntoroko district (1 if yes)	0.81(0.40)
Drought experience	Respondent has had drought experience (1 if yes)	0.96(0.20)
Flood experience	Respondent has had flood experience (1 if yes)	0.82(0.39)
Herd mobility	Respondent practices herd mobility (1 if yes)	0.55(0.50)
Livestock diversification	Respondent practices livestock diversification (1 if yes)	0.54(0.50)
Livestock migration	Respondent practices livestock migration (1 if yes)	0.83(0.38)
Selling livestock	Respondent practices selling livestock (1 if yes)	0.51(0.50)
Well construction	Respondent practices well construction (1 if yes)	0.31(0.46)
Nonfarm	Respondent engages in nonfarm enterprises (1 if yes)	0.08(0.27)
Owns boats	Owns boats (1 if yes)	0 0.11(0 0.32)
Owns fishnet	Owns fishnet (1 if yes)	0.10(0 0.30)
Local cattle produced last 1 year	Local cattle produced in the last 12 months (numbers)	69.97(58.56)

Source: Field data 2015 Figures in parentheses are standard deviations

Table 2 Use of IF only and IF and SF by pastoralists in the Rwenzori region

	Onset (%)		Cessation (%)		5-day (%)		Seasonal (%)		
	Full sample	Subsample	Full sample	Subsample	Full sample	Subsample	Full sample	Subsample	Total
IF and SF	50	84	43	73	35	59	30	50	59
IF only	40	99	38	95	33	82	24	60	41

Source: Survey data 2015 Full sample = 269, subsample IF and SF = 160, subsample IF only = 108

Table 3 Pastoralists' land tenure and access to land

Variable	Variable definition	Full sample ($N = 269$)
Customary	Customary (1 if yes)	0.34(0.47)
Freehold	Freehold (1 if yes)	0.14 (0.35)
Inheritance	Inheritance (1 if yes)	0.40(0.49)
Purchased	Purchased (1 if yes)	0.33(0.47)
Grabbed	Grabbed (1 if yes)	0.05(0.22)

Source: Survey data 2015 Figures in parentheses are standard deviations

Table 4 Barriers to climate change adaptation among pastoralists

	Mean	Std. Dev.
Inadequate or no access to extension services	0.27	0.44
Inadequate or no information on climate change	0.49	0.50
Lack of access to credit facilities	0.18	0.38
Insecure land tenure or property rights	0.09	0.29
High cost of hired labor	0.15	0.36
High cost of adaptation measures	0.26	0.44
Labor shortages	0.08	0.27
No barrier	0.04	0.20

Source: Field Data 2015

management in Africa and South America (IPCC 2018b). Climate-related risks experienced were floods and droughts. The majority (47%) had attained primary education, and 38% had no formal education. The common adaptation methods were livestock migration, herd mobility, livestock diversification and livestock sales. Enablers to adaptation included access to land through purchasing (suggesting that functional land markets existed) and inheritance (Table 3). Pastoralists diversified their livelihoods to include other natural resources such as fisheries resources, which were demonstrated by ownership of boats and fishnets (Table 1).

Barriers to Climate Change Adaptation among Pastoralists

The high cost of adaptation is a major barrier in wealthy pastoral households (Table 4). According to a key informant, it costs $800 to construct a private dam, and sinking a borehole would cost $5400 in the study area. These costs are high for smallholder pastoralists. The key informant also said that the government program, National Agricultural Advisory Services (NAADS), used to provide fence materials to a few model pastoralists. The NAADS program ended in 2014, and the cost of infrastructure development in pastoral areas is high. The fencing materials were provided to a few pasture demonstration farmers (Dalipagic and Elepu 2014; GoU 2018). The government also provided tractors for pasture improvement but not water for pastoralists (NAADS 2017). Climate finance should address such barriers. The Uganda government provided support for the construction of over 650 communal valley dams and 135 private dams for ranches in the rangelands (Nema 2001). Some of these dams dried up during droughts resulting in livestock mortality and an increase in herd mobility (Zziwa et al. 2012). Lack of hydrographic surveys for communal valley dams, poor designing of water sources without taking into account the rangeland dynamics and context, poor maintenance, and lack of community ownership were the main factors that led to the poor performance of communal valley dams (Mugerwa et al. 2014; Nema 2001). Community involvement in the management of micro-dams is higher in the Southwest and Central regions of Uganda (40–50%) than in the North and Northwest regions, where it is very low (less than 10%) (Bashar et al. 2004).

The poor performance of valley dams has increased water scarcity in the rangelands, resulting in an increase of mobile pastoralism and sociopolitical conflicts (Nema 2007), yet the government built the water sources to promote sedentary pastoralism. Even in situations where there was good performance and management of the dams, there was land degradation of the rangelands surrounding the water source (Egeru et al. 2015). Individual pastoralists tend to manage their private water sources better than they manage communal ones. The Botswana government provides support to livestock farmers with infrastructural development assistance, such as for fencing materials, and construction of boreholes to improve their adaptive capacity against droughts (Pauw 2013). Climate finance should support pastoral initiatives toward adaptation by providing fencing materials and sustainable water sources in communal rangelands and paddocks (Denton 2010).

Poor access to climate change information and poor access to extension services were also important barriers to adaptation among pastoral households. Extension workers have not mainstreamed climate change information in their dissemination of extension messages, but emphasis has been put on livestock production (AfranaaKwapong and Nkonya 2015; GoU 2016; Nkonya et al. 2015). Furthermore, in Uganda (as elsewhere in Africa) the extension system has gone through many changes from a unified extension system (1981–2000) to a demand-driven extension under the NAADS (2001–2014), to the current single-spine extension system (2015 to date) (Barungi et al. 2016; GoU 2016; Rwamigisa et al. 2018). The extension services have been tailored to sedentary but not mobile pastoralism. The demand-driven extension system under NAADS was relevant not to mobile pastoralists but to sedentary pastoralists. The extension did not serve all the pastoralists but concentrated on a few, especially model farmers (AfranaaKwapong and Nkonya 2015; GoU 2016). Furthermore, pastoralists in remote areas did not access extension services (AfranaaKwapong and Nkonya 2015). There were also a limited number of qualified service providers in rural areas, resulting in services engaging unskilled and poor quality providers (Feder et al. 2011; GoU 2016). Most of the service providers had such a narrow scope of the subject matter that they were incompetent and lacked capacity to handle other aspects such as plant and animal diseases (GoU 2016; Rwamigisa et al. 2018). There was reportedly much political interference with the implementation of demand-driven extension service (Joughin and Kjær 2010;

Rwamigisa et al. 2018). This included the use of members of the Uganda Peoples Defence Forces in the distribution of inputs and various presidential directives on the implementation of the extension services, especially during election periods (Rwamigisa et al. 2018).

Poor access to credit facilities as a barrier mostly applied to small-scale sedentary pastoralists. Large-scale pastoralists tended to finance their livestock enterprises through destocking, using rural livestock markets (Bryan et al. 2013). Furthermore, these pastoralists diversified into fishing and other nonfarm enterprises (Nkuba and Sinha 2014). The large-scale pastoralists were much more involved in livestock migration to the DR Congo as a means of adapting to climate-related risk. During severe droughts, large-scale pastoralists incur losses but recover much more quickly than small-scale pastoralists (Little et al. 2001). According to key informants, about 10,000 cattle were lost in the Ntoroko district during the severe drought of 2011. In some cases, smallholders tend to resort to agro-pastoralism in a sedentary life after high livestock mortality due to severe drought (Berhanu and Beyene 2015).

Insecure land tenure or property rights (Table 4) was another barrier to climate change adaptation in pastoralism. The rangeland policy in Uganda supports sedentary pastoralism (Byakagaba et al. 2018), while pastoralists with land enjoy dual grazing rights. During the rainy season, they graze their livestock in the paddocks. However, during pasture and water scarcity periods, such as during long dry spells and droughts, they graze communally. The areas that used to be grazed communally have been taken up by crop farmers and wealthy pastoralists who have paddocks. Over the years, "the commons" have therefore shrunk in the Rwenzori region, yet the cattle population has not decreased substantially (GoU 2006). During political instability in the DR Congo, the livestock population in rangelands of Ntotoko is much higher than its carrying capacity. The debate concerning having titled land in rangelands and communal grazing land in the cattle corridor continues. Dual grazing rights of wealthy livestock farmers who have access to large acreages have also been reported in Botswana as a coping mechanism during severe droughts (Perkins 1996).

Labor shortages and the associated high cost of labor for pastoralism (Table 4) have been caused by the labor dynamics in the study region. Most herdsmen are youth, many of whom have resorted to nonfarm activities such as motorcycle hire services (commonly called *bodaboda*) in rural areas. Furthermore, there has been a mass migration of the youth from pastoral households to urban centers and towns in search of so-called greener pastures. Thus, the high demand for herdsmen in the face of a diminished labor supply in rangelands has led to the high cost of labor. The labor of herdsmen is in high demand for such adaptation strategies as livestock migration to the DR Congo, herd mobility to River Semiliki, and pasture lands in the neighborhood of the protected areas.

Sources of the Barriers

The question that remains is whether pastoralists can overcome these barriers. As noted earlier, to answer this question, the sources of barriers needed to be identified using the barriers to adaptations diagnostic framework. The sources were then analyzed using the three structural elements in the framework, namely, the actors (referring to pastoralists), governance (referring to policies, implementation frameworks, and laws that govern adaptation or hinder it), and socio-ecological system (life in the rangeland, sometimes referred to as system of concern) (Moser and Ekstrom 2010). The three structural elements were used to examine strategies of overcoming the barriers, as is reflected in the following discussion.

The high cost of adaptation was due to the poor performance of valley dams, an insufficient number of water sources in the rangelands, policy emphasis on support for communal water sources at the expense of individual pastoralists, and equity concerns where support was only given to a few model pastoralists. This suggests that the sources of barriers were principally governance (institutional), technological, and human resource concerns. As a solution, pastoralists could be supported in infrastructure development such as fencing and construction of dams (in communal rangelands) and of boreholes in paddocks for sedentary pastoralists using climate finance from both national and international agencies. The recent change in extension policy from demand-driven extension to single-spine may also help mainstream climate change in its implementation and take into account cost of adaptation (GoU 2016). This implies an increase in allocation of resources related to climate change adaptation. The poor performance of valley dams had a bearing on the hydrological dynamics in rangelands, implying that the socio-ecological system was also a source of barrier. This can be overcome by making good use of the available expertise in hydrology to carry out feasibility studies before the water sources are constructed. Human resources in hydrology and appropriate technology for the various agro-ecological zones from the Ministry of Water and Environment should be utilized to good effect, both in monitoring and in program design of water for agricultural production facilities (GoU 1999).

The source of barriers such as poor access to climate change information and poor access to extension services was governance (institutional concerns). From a governance perspective, rangeland and extension policies were not relevant to the context. Past extension policies, both the unified extension policy, which used the training and visit system, and the demand-driven policy under NAADS, did not meet the expected outcomes (Rwamigisa et al. 2018; World Bank 2007). This has been mitigated through the new extension policy based on the single-spine extension system, which also has mainstreamed climate change information in the extension messages (GoU 2016). However, the new rangeland policy is not compatible with the socio-ecological system (Byakagaba et al. 2018). Mobile pastoralism makes good use of fragile ecosystems in the rangelands under changing climate conditions (Bailey and Brown 2011; Weber and Horst 2011). Instead, the ultimate goal of the new policy is to convert hitherto mobile pastoralists into agro-pastoralists and sedentary farmers (Wurzinger et al. 2009). There is therefore need to lobby the policy-makers to take into account both mobile and sedentary pastoralism in the rangeland policy. There is also the problem of inadequate human resources in terms of extension officers to serve the entire country. Only 35% of the technical positions in local governments had been filled in 2016 (GoU 2016). This can be overcome through increased supply and recruitment of qualified labor from tertiary agricultural institutions in the country, such as Makerere University and Bukalasa Agricultural College. Provision of incentives to extension staff is also critical to achieving sustainable outcomes.

Poor access to credit facilities is an actor-centric barrier. This can be overcome with pastoralists joining savings and credit organizations dedicated to pastoralists (Mpiira et al. 2014), probably supported by climate finance arrangements.

Insecure land tenure or property rights are governance-related barriers arising from concern related to the contradictory rangelands and land-use policies. The land policy promotes mobile pastoralism (GoU 2013) while land-use and rangeland policies discourage mobile pastoralism (GoU 2006); hence there is lack of policy coherence. This is another reason to lobby for change in the land-use and rangeland policies to take into account mobile pastoralism concerns.

The high cost of labor and labor shortages are governance-related and actor-centric barriers. From the governance perspective, there is a need for government policies that attract youth in rural areas of rangelands. The Ugandan government has passed a minimum wage law, but it is unclear how effective it will be in rangelands. From the actor perspective, pastoralists need to increase the amount of wages paid to herdsmen to attract more youth in the socio-ecological system.

Origins of the Barriers

The origins of the barriers were investigated using temporal and spatial analysis (Table 5).

Lack of credit is a contemporary and proximate barrier, implying that it is within the cycle of influence of pastoralists and a more recent barrier that can be overcome. Wealthy pastoralists finance their enterprises from destocking using rural livestock markets. Destocking is also a climate change adaptation strategy. Thus, poor access to credit mostly applies to the poor sedentary pastoralists. There has been prevalence of savings and credit organizations (SACOs) in rural areas even among pastoralists (Mpiira et al. 2014).

The high cost of hired labor is a contemporary and legacy barrier, implying it is within the cycle of influence of pastoralists and has been in play over a long time. The rural-urban migration of youth in the socio-ecological system has led to an

Table 5 Opportunities for influence and intervention to overcome the barriers

		Temporal	
Spatial Jurisdictional		Contemporary	Legacy
	Proximate	Lack of access to credit	High cost of hired labor
	Remote	Insecure land tenure, high cost of adaptation	Inadequate or no access to extension services Inadequate or no information on climate, labor shortages

Source survey data 2015. Adapted from Moser and Ekstrom (2010)

increase in the cost of labor. The youth nowadays prefer to engage in nonfarm activities such as the motorcycle hire business (popularly known locally as *bodaboda*), and move to urban centers. The promotion and implementation of universal primary education (UPE) even in pastoral areas has improved the skills of youth and provided opportunities outside the rangelands (Appleton 2001). Before the implementation of UPE, the cost of labor was very low as the supply of labor was high because many youth did not go to school. This barrier can be overcome by encouraging pastoralists to destock and by enabling them to obtain credit to finance their labor needs.

Insecure land tenure or property rights and the high cost of adaptation are contemporary and remote barriers, implying that they are a recent occurrence but outside the cycle of influence among the pastoralists due to change in government policies. Mobile pastoralism was commonly practiced in the cattle corridor in Uganda. The increase in human and livestock populations and conflicts among pastoralists and with pastoralists in the neighboring arable farming communities resulted in a change of government support, from mobile pastoralism to sedentary pastoralism. The change in rangeland land policy from mobile pastoralism to sedentary pastoralism is a result of armed conflict between arable farmers (from Teso) and pastoralists (the Karamanjong) in northeastern parts of Uganda. Before 1980, mobile pastoralism thrived in Uganda, until the Karamanjong got access to arms, which led to armed conflict in the rangelands (USAID 2011). The change in the land policy and act, with an emphasis on titling land, resulted in the shrinking of the commons in the rangelands, in turn resulting in an increase in insecure land tenure among the commoners (GoU 2013). This barrier could be mitigated through amending the land policy to provide for common grazing lands in land policy in the rangelands and to avoid titling every piece of land in Uganda. This debate has been ongoing among the various nonstate actors since the enactment of the land act in 2004 (Byakagaba et al. 2018). Blanket policy recommendations did not take into account the local setting in different parts of the cattle corridor.

The high cost of adaptation can be overcome through an increase in climate finance to include support to more pastoralists who are not model farmers, as is the current practice under the NAADS implementation framework. There has been a change in the national extension policy from the demand-driven to the single-spine NAADS, which is more inclusive and has taken into account equity concerns (GoU 2016).

Poor access to extension services and climate change information and labor shortages are legacy and remote barriers, implying that they lie outside the cycle of influence of pastoralists and have been in existence over a long period of time. Extension services should have provided the climate change information to the pastoralists, but the ineffective demand-driven extension policy had detrimental outcomes on both. This has fortunately been mitigated through the change in extension policy from the demand-driven to single-spine model (GoU 2016). The new policy promotes mainstreaming climate change information and providing services to all pastoralists.

Labor shortages are due to the failure of a minimum wage law and effective rural development strategies to address the rural–urban migration in Uganda. As observed earlier, educated youth will no longer work for very low incomes as herdsmen in pastoral areas. Recently government passed a minimum wage law whose outcomes are yet to be evaluated in rangelands. It is hoped that this will serve to retain youth in rangelands and promote urban-rural migration among the youth that have already left for urban areas. There is need for government to implement effective rural development strategies that would make rural areas more attractive to the youth. So far, strategies such as the Plan for Modernisation of Agriculture have not been effective in reversing the rural-urban migration of the youth.

It is noteworthy that the remote barriers are due to national policies such as extension and land policies that were heavily influenced by donors such the World Bank (Deininger 2003; Deinlnger and Binswanger 1999; Rwamigisa et al. 2018). Local ownership, support to home-grown initiatives, and bottom-up approaches to climate change adaptation would enhance and facilitate the development of the adaptive capacity of pastoralists against these barriers (World Bank 2007) .

Enablers of Adaptation in Uganda

Alongside barriers to climate change adaptation, studies have also identified adaptation enablers (Azhoni et al. 2018), such as human resource, institutional, governance, and economic enablers (Moser and Ekstrom 2012). In Uganda, climate change adaptation enablers include the following:

- *Human resources*, such as extension staff and climate change adaptation researchers, pastoralists with indigenous knowledge of climate change adaptation and forecasts.
- *Institutions*, such as climate change units and climate change policy, research centers for agricultural innovations, the Uganda National Meteorological Authority that provides early warning information and seasonal climate forecasts, mass media that disseminates climate and climate change information (GoU 2015).
- *Economic*, such as national and local government resource allocations to climate change adaptation initiatives, donor support from Denmark, the United States, Ireland, the United Kingdom, Belgium, Norway, the World Bank, the United Nations Development Programme, support for climate-smart agriculture from the Food and Agricultural Organisation of the United Nations, nonstate actors such as nongovernment organizations implementing climate change adaptation initiatives at local government level such as Oxfam, World Vision, Care, Volunteer Effort for Development Concerns (GoU 2015).
- *Political will* both at national and local government levels and parliamentary fora on climate change. The government provides mechanisms for flexible and negotiated cross-border access to pastoral resources under the land policy, such as access to rangelands in Eastern DR Congo (GoU 2013).

Conclusion

The study has established that the barriers to pastoralism are mainly actor-centric and governance concerns. The main barriers were poor access to climate change information, poor access to extension services, high cost of adaptation measures, poor access to credit, and insecure land tenure. Some of the policies under which pastoralism is practiced hinder adaptation to climate change. A change in the policies will enhance adaptation; for example, the recent change from the unified extension system and demand-driven extension system to the single-spine system will help in overcoming poor access to extension services and climate change information. Climate finance programs should be made more effective in addressing the high cost of adaptation. Enhancing social capital can help in overcoming poor access to credit. A change in land-use and rangeland policies will help in mitigating insecure land tenure. Coherence of land, land-use, extension, and rangeland policies will enhance pastoral adaptive capacity. It is noteworthy that there are several enablers for climate change adaptation in Uganda, which could be exploited to good effect.

The key lessons learned are that policies associated with climate change adaptation either hinder or enhance the adaptive capacity of pastoral households. Enhancing social capital such as pastoralists' savings and credit groups facilitates access to credit, which improves the capacity to overcome some barriers. Climate finance should be administered in an equitable manner to improve the adaptive capacity for both wealthy and poor households in rural areas.

Future prospects for overcoming barriers in developing countries lie in the identification of the origins and sources of the barriers for effective policy interventions. Policy coherence in climate change adaptation implementation is critical to achieving sustainable development goals, especially Goals 13 and 15, which address the problems of climate change and terrestrial ecosystems, respectively. Mainstreaming climate change in, and effective implementation of, rural development policies will improve the ability of pastoralists to overcome the barriers to climate change adaptation. Rangeland policies should promote both mobile and sedentary pastoralism.

Acknowledgments The authors are grateful for the support during the research from research assistants and the respondents who made this research a reality through their favorable cooperation. This research was self-financed by the lead author for data collection, and special thanks to Nassali Mercy Nkuba for moral support. We are grateful to Kabarole Research and Resource Centre for their collaboration during the field work and the Government of Uganda for permission to carry out the study. Special thanks to Susane Moser and Julia Ekstrom for the insights in the diagnostic framework. This research is part of PhD work for Michael Robert Nkuba, who wishes to thank the University of Botswana for waiving tuition costs.

References

AfranaaKwapong N, Nkonya E (2015) Agricultural extension reforms and development in Uganda. J Agric Extd Ural Dev 7(4):122–134

Alam GM, Alam K, Mushtaq S, Khatun MN, Leal Filho W (2018) Strategies and barriers to adaptation of hazard-prone rural households in Bangladesh Limits to Climate Change Adaptation. Springer, Swirtezland, (pp. 11–24)

Antwi-Agyei P, Dougill AJ, Stringer LC (2015) Barriers to climate change adaptation: evidence from Northeast Ghana in the context of a systematic literature review. Clim Dev 7(4):297–309
Appleton S (2001) What can we expect from universal primary education? In: Uganda's recovery: the role of farms, firms, and government. The World Bank, Washington, DC, pp 371–406
Azhoni A, Jude S, Holman I (2018) Adapting to climate change by water management organisations: enablers and barriers. J Hydrol 559:736–748
Bailey DW, Brown JR (2011) Rotational grazing systems and livestock grazing behavior in shrub-dominated semi-arid and arid rangelands. Rangel Ecol Manage 64(1):1–9
Barungi M, Guloba M, Adong A (2016) Uganda's Agricultural Extension Systems: How appropriate is the Single Spine Structure? Research report No 15. Economic Policy Research Centre. Makerere University-Kampala
Bashar K, Kizaa M, Chane B, Seleshi Y (2004) Assessment of existing micro dams in Ethiopia, Sudan and Uganda. Micro-dam Group Nile Basin Capacity Building Network River Engineering. River Structure Research Cluster. Adis Ababa
Berhanu W, Beyene F (2015) Climate variability and household adaptation strategies in Southern Ethiopia. Sustainability 7(6):6353–6375
Bryan E, Ringler C, Okoba B, Roncoli C, Silvestri S, Herrero M (2013) Adapting agriculture to climate change in Kenya: household strategies and determinants. J Environ Manage 114:26–35. https://doi.org/10.1016/j.jenvman.2012.10.036
Byakagaba P, Egeru A, Barasa B, Briske DD (2018) Uganda's rangeland policy: intentions, consequences and opportunities. Pastoralism 8(1):7
Cochran WG (1963) Sampling Technique. (Vol. 2nd Edition,). New York.: John Wiley and Sons Inc
Dalipagic I, Elepu G (2014) Agricultural value chain analysis in northern Uganda: Maize, rice, groundnuts, sunflower and sesame. Against Hunger (ACF) International. Kampala, Uganda
Deininger KW (2003) Land policies for growth and poverty reduction. World Bank Publications, Washington DC
Deinlnger K, Binswanger H (1999) The evolution of the World Bank's land policy: principles, experience, and future challenges. World Bank Res Obs 14(2):247–276
Denton F (2010) Financing adaptation in least developed countries in West Africa: is finance the 'real deal'? Clim Pol 10(6):655–671
Egeru A, Wasonga O, MacOpiyo L, Mburu J, Tabuti JR, Majaliwa MG (2015) Piospheric influence on forage species composition and abundance in semi-arid Karamoja sub-region, Uganda. Pastoralism 5(1):12
Feder G, Birner R, Anderson JR (2011) The private sector's role in agricultural extension systems: potential and limitations. J Agribus Dev Emerg Econ 1(1):31–54
Filho WL, Nalau J (2018) Introduction: limits to adaptation. Springer, Cham
Government of Uganda (GoU) (1999) National water Policy. Ministry of Water and Environment. Kampala, Uganda
Government of Uganda (GoU) (2006) The Uganda National landuse Policy. Ministry of Land Housing and Urban Devlopment Kampala, Uganda
Government of Uganda (GoU) (2013) The Uganda National land Policy. Ministry of Land Housing and Urban Devlopment Kampala, Uganda
Government of Uganda (GoU) (2015) National climate change Policy. Ministry of Water and Environment. Kampala, Uganda
Government of Uganda (GoU) (2016) National Agricultural Extension Policy. Ministry of Agriculture, Animal industry and Fisheries. Entebbe, Uganda
Government of Uganda (GoU) (2018) Guidelines For Use Of Production And Marketing Grant By Local Governments For Fy 2018/19. Ministry of Agriculture, Animal industry and Fisheries. Entebbe, Uganda
Greenough KM (2018) Pastoralists shifting strategies and perceptions of risk: post-crisis recovery in Damergou, Niger. In: Limits to climate change adaptation. Springer, Switerzland, pp 129–142
IPCC (2014a) Climate change 2014: adaptation opportunities, constraints, and limits. Retrieved from Cambridge University Press, Cambridge, UK/New York

IPCC (2014b) Climate change 2014: impacts, adaptation, and vulnerability. Part B: regional aspects. Retrieved from Cambridge University Press, Cambridge, UK/New York

IPCC (2018a) Impacts of 1.5 °C of Global Warming on Natural and Human Systems. Retrieved from Cambridge University Press, Cambridge, UK/New York

IPCC (2018b) Strengthening and implementing the global response. Retrieved from Cambridge University Press, Cambridge, UK/New York

IPCC (2018c) Sustainable development, poverty eradication and reducing inequalities. Retrieved from Cambridge University Press, Cambridge, UK/New York

Joughin J, Kjær AM (2010) The politics of agricultural policy reform: the case of Uganda. Forum Dev Stud 37(1):61–78

Juana JS, Okurut FN, Makepe PM, Kahaka Z (2016) Climate change perceptions and adaptations for livestock farmers in Botswana. Int J Econ Issues 9(1):1–21

KRC & RFPJ (2012) Stuck in the Mist: Contextual Analysis of the Conflicts in the Rwenzori Region. KRC (Kabarole Research and Resource Centre) and RFPJ (Rwenzori Forum for Peace and Justice). Fort Portal, Uganda

Little PD, Smith K, Cellarius BA, Coppock DL, Barrett C (2001) Avoiding disaster: diversification and risk management among East African herders. Dev Chang 32(3):401–433

Little PD, McPeak J, Barrett CB, Kristjanson P (2008) Challenging orthodoxies: understanding poverty in pastoral areas of East Africa. Dev Change 39(4):587–611

Löf A (2013) Examining limits and barriers to climate change adaptation in an Indigenous reindeer herding community. Clim Dev 5(4):328–339

Luseno WK, Mcpeak JG, Barrett CB, Little PD, Gebru G (2003) Assessing the value of climate forecast information for pastoralists: evidence from Southern Ethiopia and Northern Kenya. World Dev 31(9):1477–1494. https://doi.org/10.1016/S0305-750X(03)00113-X

Lybbert TJ, Barrett CB, Mcpeak JG, Luseno WK (2007) Bayesian herders: updating of rainfall beliefs in response to external forecasts. World Dev 35(3):480–497. https://doi.org/10.1016/j.worlddev.2006.04.004

Moser SC, Ekstrom JA (2010) A framework to diagnose barriers to climate change adaptation. Proc Natl Acad Sci 107(51):22,026–22,031

Moser SC, Ekstrom JA (2012) Identifying and overcoming barriers to climate change adaptation in San Francisco Bay. California Energy Commission. Sacramento, CA 95814, USA

Mpiira S, Kiiza B, Katungi E, Tabuti J, Staver C, Tushemereirwe W (2014) Determinants of net savings deposits held in savings and credit cooperatives (SACCOs) in Uganda. J Econ Int Financ 6(4):69–79

Mugerwa S, Kayiwa S, Egeru A (2014) Status of livestock water sources in Karamoja sub-region, Uganda. Resour Environ *4*(1):58–66

Muller C, Shackleton SE (2014) Perceptions of climate change and barriers to adaptation amongst commonage and commercial livestock farmers in the semi-arid Eastern Cape Karoo. Afr J Range Forage Sci 31(1):1–12

NAADS (2017) NAADS Strategic Plan 2015/16 – 2019/20. Ministry of Agriculture, Animal industry and Fisheries. Entebbe, Uganda

NAPA (2007) Uganda National Adaptation Plans of Action. Ministry of Water and Environment. Kampala, Uganda

NEMA (2001) State of the Environment Report for Uganda 2001. National Environment Management Authority (NEMA), Kampala. Uganda

NEMA (2007) State of the Environment Report for Uganda 2007. National Environment Management Authority (NEMA), Kampala, Uganda

Nkonya E, Place F, Kato E, Mwanjololo M (2015) Climate risk management through sustainable land management in Sub-Saharan Africa. In Sustainable Intensification to Advance Food Security and Enhance Climate Resilience in Africa. Springer, Switzerland, (pp. 75–111)

Nkuba MR, Sinha N (2014) Aquaculture and fishers' livelihood diversification in Uganda –an empirical analysis. Paper presented at the IASTED international conference environment and water resource management (AfricaEWRM 2014), Gaborone

Pauw P (2013) The role of perception in subsistence farmer adaptation in Africa: enriching the climate finance debate. Int J Clim Change Strategies Manage 5(3):267–284

Perkins J (1996) Botswana: fencing out the equity issue. Cattleposts and cattle ranching in the Kalahari Desert. J Arid Environ 33(4):503–517

Rwamigisa PB, Birner R, Mangheni MN, Semana A (2018) How to promote institutional reforms in the agricultural sector? A case study of Uganda's National Agricultural Advisory Services (NAADS). Dev Policy Rev 36(5):607–627

Shackleton S, Ziervogel G, Sallu S, Gill T, Tschakert P (2015) Why is socially-just climate change adaptation in sub-Saharan Africa so challenging? A review of barriers identified from empirical cases. Wiley Interdiscip Rev Clim Change 6(3):321–344

Speranza CI, Kiteme B, Ambenje P, UrsWiesmann, Makali S (2010) Indigenous knowledge related to climate variability and change: insights from droughts in semi-arid areas of former Makueni District, Kenya. Clim Change 100:295–315. https://doi.org/10.1007/s10584-009-9713-0

USAID (2011) Climate Change and Conflict in Uganda: The Cattle Corridor and Karamoja. CMM Discussion Paper No 3 USAID. Kampala, Uganda

Weber KT, Horst S (2011) Desertification and livestock grazing: the roles of sedentarization, mobility and rest. Pastoral Res Policy Pract 1(1):19

World Bank (2007) World Development report 2008. Agriculture for Development. World Bank. Washington DC, USA

Wurzinger M, Ndumu D, Baumung R, Drucker A, Okeyo A, Semambo D et al (2006) Comparison of production systems and selection criteria of Ankole cattle by breeders in Burundi, Rwanda, Tanzania and Uganda. Tropl Anim Health Prod *38*(7):571–581

Wurzinger M, Okeyo AM, Semambo D, Souml J (2009) The sedentarisation process of the Bahima in Uganda: an emic view. Afr J Agric Res 4(11):1154–1158

Zziwa E, Kironchi G, Gachene C, Mugerwa S, Mpairwe D (2012) The dynamics of land use and land cover change in Nakasongola district. J Biodivers Environ Sci *2*(5):61–73

6

Climate Change Adaptation Mechanism for Sustainable Development Goal 1 in Nigeria: Legal Imperative

Erimma Gloria Orie

Contents

Abstract

Despite international efforts on poverty reduction in the last decade, poverty is rampant in many countries including Nigeria. Poverty remains a principal challenge for development in twenty-first century and a threat to achievement of Sustainable Development Goal (SDG) 1, which is a global attempt, among others, to end poverty by 2030. Meanwhile, 13 out of the 15 countries where extreme poverty is rising are in Africa. According to the World Poverty Clock, Nigeria, by 2018, had the largest extreme poverty population of 86.9 million, thus making the people vulnerable to malnutrition, armed conflict, migration, and other socioeconomic and environmental shocks. Whereas these impacts are exacerbated by climate change (CC), unfortunately, Nigeria's adaptation efforts are inadequate due to certain impediments. The chapter finds that Nigeria lacks the CC law to properly regulate institutional and policy interventions to impacts of CC. It argues that although adaptation as opposed to mitigation is interim, yet integrating adaptation measures into Sustainable Development (SD) framework and poverty reduction strategies is a potent means of addressing CC impacts on the poor and achieve SDG1 target. The chapter therefore recommends the establishment of CC

E. G. Orie (✉)
Department of Private and Property Law, National Open University of Nigeria, Abuja, Nigeria

law in Nigeria that incorporates adaptation measures in poverty reduction strategies and mainstreaming of CC issues.

Keywords

Poverty · Adaptation · Sustainable development (SD) · Sustainable Development Goals (SDGs) · Climate change (CC)

Introduction

Since the turn of the twenty-first century, there has been an unrivaled concern of the international community to achieve sustainable development (SD). From Stockholm Declaration through Rio +10, Agenda 21 to the United Nations Conference on Sustainable Development held in Rio 2012, and the Millennium Development Goals (MDGs), the Paris Agreement, and the recent UN Conference of Parties (COP 24) at Bon, Germany, the focus has been to combat environmental damage and reduce poverty and diseases through global cooperation on common interests, mutual needs, and shared responsibilities. Even the United Nations 2030 Agenda for Sustainable Development is an action plan focused on people, planet, prosperity, peace, and partnership (5Ps) with a mandate that no one should be left behind. The first Sustainable Development Goals' Report published in 2016 (Portugal 2016) reveals that about 1 in every 8 persons in the world still lives in extreme poverty; some 800 million people are suffering from hunger; the birth of nearly 1 in 4 children under the age of 5 years is still unregistered; women spend about 2.4 times more hours per day on caregiving and household tasks than men; 1.1 billion people live without electricity; and that water scarcity now affects more than 2 billion people in the world (Ibid). Furthermore, the UN has launched the 2018 version of the yearly Sustainable Development Goals (SDGs) Report (SDG Report 2018) which finds that conflict, climate change, and inequality are major factors in growing hunger and displacement, and are hindering progress toward the SDGs. The implication of these reports is that the world, particularly Africa and of course, Nigeria, is not on course to ending poverty at least by 2030. This could undermine the attainment of SDG 1 which deals with poverty reduction.

The major objective of the World Bank since the 1970s has equally been to reduce poverty. In 2013, the World Bank Group announced two all-embracing goals: the end of chronic extreme poverty by 2030; the promotion of shared prosperity, defined in terms of economic growth of the poorest segments of society (to reduce extreme poverty to 3% of the world population by 2030 and, for the first time, including a distributional goal, "share prosperity" by promoting the income growth of the poorest 40%). However, the global proportion of people living in extreme poverty is three times lower than in 1970; it fell from 43% in 1990 to 19% in 2010 and further to 17% in 2011 (Ending Extreme Poverty 2013).

Despite commitments to inclusive, pro-poor, and broad-based growth, the poorest 20% of people still receive just 1% of global income causing the gap between the poorest 20% and everyone else to continue to widen. Indeed, 13 out of the 15 countries where extreme poverty is rising are in Africa. According to the World Poverty Clock, by 2018, Nigeria had the largest extreme poverty population of 86.9 million due to drought, flood, heatwaves, ocean surges, reduction in crop yield, decreased fresh water availability, disruption of economic activities, destruction of infrastructure, etc. These challenges are exacerbated by the impact of climate change (Orie 2015), the end state of which is that the people are made vulnerable to malnutrition, armed conflict, migration, and other socioeconomic and environmental shocks. The National Bureau of Statistics figures showed that poverty incidence worsened between 2004 and 2010, despite impressive growth record over this period (Olofin et al. 2015). Poor farmers also risk losing crops as the flood season occurs when crops ripen for harvest. In the longer term, poor households also risk losing wage opportunities as the sick and injured cannot work or as the disaster destroys the need for labor. Recovery strategies, like selling assets, can leave the poor without income and thus more vulnerable.

Nigeria's extreme poverty population of 86.9 million can be scary. Specifically, for Lagos, Osun, Anambra, Ekiti, Edo, Imo, Abia, and Rivers States, the poverty level hovers between 8.5 and 21.1%; Plateau, Nassarawa, Ebonyi, Kaduna, Adamawa, and Benue States have poverty level that hovers between 51.6 and 59.2%, while the poverty level for Niger, Borno, Kano, Gombe, Taraba, Katsina, Sokoto, Kebbi, Bauchi, Jigawa, Yobe, and Zamfara States hovers between 61.2 and 91.9% (Olawale 2018). Certainly, these degrees of poverty make the people vulnerable to malnutrition, armed conflict, migration, and other socioeconomic and environmental shocks. Whereas these impacts are exacerbated by climate change, unfortunately, Nigeria's adaptation efforts to counter poverty are inadequate due to certain legal impediments. Nigeria, for instance, lacks the climate change law to properly regulate institutional and policy interventions to impacts of climate change. The crux of the matter is that without a legally binding framework, integrating adaptation measures into SD framework and poverty reduction strategies, it will be difficult to address climate change impacts on the poor and achieve SDG1 target. Moreover, the commitment to all-inclusive zero poverty has no global or national framework of measurement. These are some of the challenges Nigeria has been grappling with and must overcome in order to attain SDG 1 target. Perhaps, leveraging international best practices, the establishment of climate change law in Nigeria that incorporates adaptation measures in poverty reduction strategies could present a way forward and that is the motivation and what this chapter purposes to achieve. The chapter is limited mostly to document analyses of the relevant UN and other international conventions/documentations as well as national policies on the subject matter in the last decade when climate change, MDG, and SDG issues assumed more global attention and priority. The focus of this work remains the legal dimensions to Nigeria's efforts regarding climate change adaptation mechanism.

Therefore, the chapter begins with conceptual clarifications of key words to set the work in proper perspective leading to establishment of nexus between climate change adaptation and poverty. This is followed by a highlight of global efforts to combat poverty through Adaptation mechanism. Next is an examination of the policy, legal, and institutional framework for fighting poverty in Nigeria leading to a discussion on the legal issues, challenges, as well as strategies in the application of climate change adaptation mechanism to achieve SDG 1. Thereafter the chapter concludes with salient recommendations.

Conceptual Clarification

It is necessary to conceptualize the key words in order to put the paper in proper perspective and to aid understanding. The words and phrases to be conceptualized are climate change, sustainable development, adaptation, and poverty.

Climate Change

The United Nations Framework Convention on Climate Change (UNFCCC) defines climate change as a change of climate which is attributed directly or indirectly to human activity that alters the composition of the global atmosphere and which is in addition to natural climate variability observed over comparable time periods (UNFCCC Art 1.2). The impact of climate change include rise in sea levels, rise in temperature, and change in weather patterns resulting in decline in overall biodiversity, implications for agriculture and food security, human health, unpredictable rain pattern and floods, prolonged drought and subsequent crop failures, water shortages, and homelessness (Orie 2017).

Sustainable Development

Sustainable development is the universally adopted plan to promote prosperity and social well-being while protecting the environment. The 2030 Agenda for SD is "our shared vision of humanity and a social contract between the world's leaders and the people" (Ban Ki-Moon), which sets an ambitious but achievable, universal, and holistic agenda for all. As a universal agenda, based on 17 SDGs and 169 targets to be implemented by all countries, the 2030 Agenda calls for the integration of the SDGs into the policies, procedures, and actions developed at the national, regional, and global levels. There is a clear call for action to dramatically scale up development finance and improve the development impact of all financial flows. But its goals are at risk of not being realized as the gap between the poorest people and the rest of the world continues to widen.

In several ways, the SDGs brought a novel tactic to the manner in which the world approaches the subject matter of development. These include, for example:

(i) Integrating the three dimensions of SD (economic, social, and environmental).
(ii) Being based on universal goals and targets to be implemented by all countries (and not only by developing countries).
(iii) Having a greater potential for tackling inequality and promoting human rights as a cross-cutting concern across all SDGs.

(iv) Involving new dynamic concerted efforts from a wide range of actors, including nongovernmental organizations (NGOs), the private corporate sector, academia, social partners, and other members of civil society, including the cooperation between parliament, government, regional, and local authorities. In fact, this is a challenge which concerns us all (Portugal).

Adaptation

The term adaptation according to the UNFCCC refers to "adjustments in ecological-social-economic systems in response to actual or expected climatic stimuli, their effects or impacts, and the building of a climate-resilient society that is able to withstand or recover quickly from difficult conditions caused by the adverse effects of climate change" (Medugu et al. 2011). Examples of adaptations include building flood defenses, planning for heat waves and higher temperatures, and installing water-permeable pavements to better deal with floods and storm water and improving water storage and use (Ibid; Orie 2015). Adaptation also encompasses making the most of any potential beneficial opportunities associated with climate change (for example, longer growing seasons or increased yields in some regions) (global climate change). However, the major challenge to adaptation includes the fact that adaptation is fundamentally about climate change risk management, and it is not a long-term solution, hence the need for reinforcement which mitigation offers (IPCC report).

A major benefit of adaptation is the fact that it is usually imbued with the prospects of reducing the negative impacts of climate change. However, the capacity to adapt is principally a factor of the socio-economic characteristics, and the likelihood of adaptation needs to be carefully analyzed given the peculiar circumstances of the developing countries and their obvious disadvantage with respect to technological, financial, and institutional provisions (McGuigan et al. 2002).

Poverty

The World Bank defines extreme poverty as living on less than 1.90 international dollars (int.$) (Roser and Ortiz-Ospina 2019). Poverty is the result of economic, political, and social processes that interact and frequently reinforce each other in ways that exacerbate the deprivation in which poor people live (Ibid). The United Nations has also now declared the eradication of poverty by 2030 as a primary development goal. Poverty is at the same time stark and conceptually elusive, especially when we try to track it statistically across nations and over time. The World Bank Group has been at the vanguard of developing measures of living standards and it sets the International Poverty Line (IPL). The poverty line was revised in 2015 and since then persons are considered to be in extreme poverty if they live on less than 1.90 (int.-$) per day. The measurement of this poverty is based on the monetary value of a person's consumption while the application of income measures is employed for countries without available reliable consumption measures (World Bank 2015). Alternative starting points for measuring welfare include subjective views (e.g., self-reported life satisfaction), basic needs (e.g., caloric requirements), capabilities (e.g., access to education), and minimum rights (e.g., human rights) (Ibid). There is also the national accounts method to estimate poverty which is based on academic studies that reconstruct historical income levels from cross-country macro estimates on economic output and inequality. Therefore, while the IPL is useful for understanding the changes in living conditions of the very poorest of the world, one must also take into account higher poverty lines reflecting the fact that living conditions at higher thresholds or well above the IPL can still constitute a destitute.

Nexus Between CC Adaptation Mechanism and Poverty (SDG 1)

The 2007 and 2014 IPCC reports state that due to the effect of climate change, both distributional impacts of increased temperature and precipitation changes will vary from region to region, affecting higher and lower latitudes disparagingly and asymmetrically (IPCC Fourth Assessment Report). The IPCC Report predicts further that global warming resulting from carbon emissions will cause a rise in sea levels and may lead to an increase in the frequency and severity of natural disasters. In addition, incidences of natural disasters such as droughts and floods, which affect agricultural production, fisheries and marine life, water resource availability, industry, and human health will continue to be on the increase and indeed exacerbated by climate change.

Coastal areas are most at risk to the changes outlined above. Increased sea levels will bring salinization and an intrusion of seawater into freshwater sources, flooding and loss of land, erosion, loss of wetlands and mangroves, and loss of soil fertility. Furthermore, changes in temperature will alter ocean circulation patterns, vertical mixing of water and wave patterns which will impact on marine productivity, availability of nutrients, and disturb the structure of marine and coastal ecosystems (IPCC 2001). While there is no data on the frequency of disasters due to climate change, Messer estimates that with a 70 cm sea level rise, the number of people at risk of annual flooding could increase from 46 million to 90 million. With such disasters come the urge need for interim relief assistance (like artificial irrigation in drought areas) to enable the people adjust to the impact of the disaster. Thus, adaptation enables the victims to access immediate and or interim assistance as opposed to a long-term relief which mitigation provides (Messer 2003).

Furthermore, the disasters are expected to increase the disparity in wealth between the developed and developing world, and redistributive impact is one of the major reasons for concern about the climate change phenomena as expressed by the IPCC in its 2001 report. Due to these differential effects, developing countries like Nigeria are likely to suffer more from the economic impacts of climate change, as well as being the least able to adapt to new climatic conditions. Moreover, poorer communities also have limited means to cope with the losses and damage inflicted by natural disasters. Lack of adaptation option like insurance policies, savings, or credit make it almost impossible to replace or compensate for the numerous things lost or destroyed, including houses, livestock, food reserves, household items, and tools (IPCC 2001), which in turn will threaten the chances of effective reduction in poverty level.

Developing countries often do not have the resources for these and consequently are ill-prepared in terms of coastal protection, early warning and disaster response systems, and victim relief and recovery assistance, hence the disproportionate level of climatic impact. For example, it has been asserted in some quarters that even when developed countries experience a larger proportion of property damage (75%), that recovery costs are higher for developing countries. Also, that whilst developed countries pay 0.1% of GDP in losses, developing countries pay 2–3%, or sometimes as much as 15% as seen with hurricanes in the Caribbean (Gurtner 2010). Developing countries also experience greater loss of life, about 90% of all deaths (Bankoff 1999). For example, the risk of drowning in Fiji due to dyke failure is one in 100,000, whereas in the Netherlands it is one in ten million (Olsthoorn et al. 1999). Notwithstanding the degree of risks exacerbated by climate change for these poor and marginalized populations, lots of uncertainties exist for developing countries regarding scientific predictions, levels of vulnerability, and the ability to adapt to these developments. Recent estimates suggest that, in the absence of adaptation, climate change could result in a loss of between 2% and 11% of Nigeria's GDP by 2020, rising to between 6% and 30% by the year 2050. This loss is equivalent to between N15 trillion (US$100 billion) and N69 trillion (US$460 billion) (National adaptation strategy). Several legal barriers, poor access to location, lack of services and infrastructure, and poor building structures all increase the vulnerability to flooding, storm surges, drought, and rain. The result is the accentuation of poverty, hunger, lack of quality education, gender inequality, loss of basic shelter, and jobs. These challenges are a threat to both the global and national efforts toward the attainment of the SDGs especially that of eradicating poverty.

The SDGs comprise 17 goals spliced into 169 targets. For example, the focus of SDG #1 is to end poverty and SDG #2 is to end hunger. Poverty breads hunger which in turn affects the general well-being of a person by reducing the coping capacity to environmental challenges. A poor man is usually hungry and thus cannot achieve SDG #3 which is on well-being. When the stomach is empty, nothing can enter the brain, but if the first three goals are achieved, one can make use of SDG #4 Quality Education and learn to respect others and internalize SDG #5 Gender Equality. Thus, the SDGs are interrelated and interconnected.

It has been noted that giving priority to SD and meeting the SDGs is in tandem with efforts to adapt to climate change. When SD promotes livelihood security, it enhances the adaptive capacities of vulnerable communities and households. Examples include SDG 2 and its targets that promote adaptation in agricultural and food systems (Lipper et al. 2014) and SDG 11 which supports adaptation in cities to reduce harm from disasters (Kelman 2017). The overall success of SDGs 2 and 11 will assist in reducing poverty which is SDG 1. Indeed it has been submitted by various scholars that a well-integrated adaptation supports sustainable development (Weisser 2014). Substantial synergies are observed in the agricultural and health sectors and in ecosystem-based adaptations. However, a particular adaptation strategy can equally lead to adverse consequences for developmental outcomes.

Furthermore, adaptation strategies that advance one SDG can result in tradeoffs with other SDGs. For instance, agricultural adaptation to enhance food security (SDG 2) causes negative impacts on health, equality, and healthy ecosystems (SDGs 3, 5, 6, 10, 14 and 15). Notwithstanding this important role that adaptation could play to ensure that achievement of SDG 1 remains on course, Nigeria has no climate change law for the implementation of adaptation policy to achieve SDG 1. This is quite a challenge that must be addressed in order to make progress.

This segment has so far established the nexus between climate change adaptation and poverty. It has argued that incidences of natural disasters such as droughts, floods desertification, etc. which continue to increase and indeed are exacerbated by climate change can be checkmated, albeit, in the interim by adaptation measures to ensure the eradication of poverty (attainment of SDG 1) by 2030. In the case of Nigeria, it is therefore imperative that her adaptation measures are strengthened legally to ensure that the stated target is achieved. The next section will examine the global perspective to this.

Global Efforts to Combat Poverty Through Adaptation Mechanism

There is increasing global efforts to address poverty based on our interconnectedness as peoples living on mother earth and the understanding of the need to save the environment for use by future generations. From around 1820, there was widespread poverty with more than a billion people that lived in extreme poverty, but this has changed over the last two centuries due mainly to economic growth not withstanding that the population increased sevenfold over the same period. According to these household surveys, 44% of the world population lived in absolute poverty in 1981, whereas in 32 years, the share of people living in extreme poverty was divided by 4, reaching levels below 11% in 2013. There were 2.2 billion people living in extreme poverty in 1970, while people living in extreme poverty in 2015 were 705 million (Roser and Ortiz-Ospina 2019).

The global incidence of extreme poverty has gone down from almost 100% in the nineteenth century to 10.7% in 2013. Some analysts have submitted that this substantial reduction of global poverty is not unconnected with the poverty reduction in China where in 1981 almost one third (29%) of the non-Chinese world population was living in extreme poverty, but by 2013, this share had fallen to 12%. Others believe it has to do with sustained economic growth, driven by industrial development and their ability to benefit from globalization. While this is a great achievement, there is absolutely no reason to be complacent: a poverty rate of 10.7% means a total poverty headcount of about 746 million people (Ibid). The breakdown by continent is as follows: 383 million in Africa, 327 million in Asia, 19 million in South America, 13 million in North America, 2.5 million in Oceania, and 0.7 million in Europe. Africa is the continent with the largest number of people living in extreme poverty. On the other hand, India is the country with the largest number of people living in extreme poverty (218 million people), while Nigeria and the Congo (DRC) follow with 86 and 55 million people, respectively (Ibid).

In line with the global fight against poverty, it has been suggested in some quarters that the most direct method to measure poverty is to use the poverty headcount ratio. This method requires the setting up of a poverty line and then counting the number of people living with incomes or consumption levels below that poverty line. A major advantage of this system is that it offers information that is candid to interpret; by definition, it shows the percentage of the population living with consumption (or incomes) below some minimum level (Ibid).

The World Bank Group has also published a new set of poverty estimates, as part of their report on poverty and shared prosperity. These estimates, explained in details in two related background papers (Ibid), are consistent with the official World Bank poverty figures published in Povcal and the World Development Indicators, but they are disaggregated by key demographic characteristics such as age and educational attainment. Furthermore, the UN projection through the SDGs is an extremely ambitious target of ending extreme poverty by 2030, a target which many analysts have observed is likely to require growth with declining inequality. Globally, many governments tend to adopt short-term interventions programs as a means of reducing poverty. In fact, a multilayered program offering short-term support along various household dimensions has been shown to cause lasting progress for the very poor in six different countries. This intervention comprises six elements; a productive asset grant, temporary cash consumption support, technical skills training, high frequency home visits, a savings program, as well as health education and services.

Regionally, the European Union in a show of commitment to the SD across borders published, through the European Commission on November 22, 2016, a Communication on the "*Next steps for a sustainable European future*" that sets out how the 2030 Agenda is to be implemented within the EU. This internal implementation of the Agenda includes two work streams: the first is to fully integrate the SDGs in the European policy framework and the Juncker Commission's ten priorities for its current term, identifying the most relevant sustainability concerns, and also assessing European policies and the efforts to achieve the 17 Goals, while the second one will work on further developing a long-term European vision and focus on sector-based policies after 2020 which will enable the long-term implementation of the SDGs. The new Multiannual Financial Framework beyond 2020 shall also reorient the EU budget's contributions toward that same end (Ibid). Furthermore, for the external implementation of the 2030 Agenda, the Commission also presented on November 22, 2016 a communication on the review of the European Consensus on Development (2017), seeking to adapt the development policy of the EU to the new international development architecture. European Consensus was adopted in 2017, and it is organized around the 5Ps of the Agenda 2030. It is expected to impact on the elaboration of development instruments and programs of the EU and its Member States, fostering their alignment with the SDGs and the Addis Ababa Action Agenda (Addis Ababa 2015).

The implementation of the Agenda at the national level brings new challenges which require some reshaping of institutional models to reflect and meet the inherent cross-sector coordination requirements. In this process, it is also relevant to create mechanisms that provide the necessary coordination between the various institutional stakeholders, with a view to present progress reports in a number of different fora in which the implementation of the 2030 Agenda is discussed. On the other hand, the EU's impact outside its borders is not limited to its external action agenda. Many EU policies contribute to the implementation of the SDGs worldwide. Therefore, achieving coherence across all EU policies is crucial to achieving the SDGs. With respect to bilateral and multilateral donor programs, the centerpiece of USAID's climate change policy is the Climate Change Initiative (CCI) with projects in three target areas, reducing GHGs, increasing developing countries' participation in UNFCCC, and decreasing vulnerability. The Canadian Government also established the Climate Change Development Fund (CCDF) in 2000, which allocates US$100 million to assisting developing countries combat the causes and effects of climate change. The CCDF has four priority areas, namely emissions reduction, carbon sequestration, adaptation, and capacity building. Key recipients for the emission reduction include Brazil, China, Egypt, India, Kazakhstan, Nigeria, and South Africa (37–42% of funds), while the key recipients for adaptation are Bangladesh, sub-Saharan Africa, and Small Island States (10–15% of funds) (Climate change 2016). Thus, for Nigeria to remain a beneficiary of these foreign aids and attract foreign investment to complement her efforts, she must align her policy frameworks to SDGs to meet international standards in terms of harmonization of policies and legality of initiatives while keying into the priority project target areas set by donor countries and agencies.

In sum, poverty-induced climate change is widespread and has become a global phenomenon which requires international commitments for it to be curbed through adaptation mechanism. The aforementioned efforts have provided the global community with a template for the implementation of adaptation mechanism. In all of these, it will be interesting to examine how Nigeria has fared in employing adaptation mechanism, and this will form the subject of discussion for subsequent segments of this chapter.

Policy, Legal, and Institutional Frameworks for Fighting Poverty in Nigeria

Nigeria can boast of numerous policies, legal, and institutional frameworks which directly and indirectly deal with climate change and poverty reduction. The frameworks for alleviating poverty in Nigeria include:

(i) Nigeria's Constitution, particularly its Chaps. II and IV on the protection of citizen's right and the Directive principles of state policy.
(ii) Nigeria Climate Change Policy Response and Strategy: The focus of the policy is to:
    (a) Implement mitigation measures that will promote low carbon as well as sustainable and high economic growth
    (b) Enhance national capacity to adapt to climate change

(c) Strengthen national institutions and mechanisms (policy, legislative, and economic) to establish a suitable and functional framework for climate change governance

(iii) The National Adaptation Strategy and Plan of Action on Climate Change for Nigeria (NASPA-CCN). This is the principal National Adaptation Strategy and Plan of Action on Climate Change for Nigeria. The objective of the policy is to integrate climate change adaptation into national, sectoral, state, and local government planning and into the plans of universities, research, and educational organizations, civil society organizations, the private sector, and the media; mobilize communities for climate change adaptation actions through the provision of appropriate user-friendly information; and reduce the impacts of climate change on key sectors and vulnerable communities, among others. Specifically, with respect to agriculture (crops and livestock), the policy seeks to adopt improved agricultural systems for both crops and livestock and to provide early warning/meteorological forecasts and related information.

(iv) The National Policy on Environment which supports "the prevention and management of natural disasters such as floods, drought, and desertification."

(v) Nigeria's Agricultural Policy whose objectives include the protection of "agricultural land resources from drought, desert encroachment, soil erosion, and floods."

(vi) The Nationally Determined Contributions (NDC) which is a determined contribution of Nigeria as regards her target of carbon emission reduction.

(vii) Nigeria's Drought Preparedness Plan.

(viii) National Policy on Erosion and Flood Control.

(ix) National Water Policy, National Forest Policy.

(x) National Health Policy.

(xi) Nigeria's Sovereign Green Bond – Nigeria embraced the issuance of its Sovereign Green Bonds as an innovative and alternate way of raising finance both locally and internationally. This is a financing mechanism to facilitate and assist Nigeria meet its Nationally Determined Contribution target and low carbon pathway for socio-economic development in line with the Economic Recovery Growth Plan (ERGP).

(xii) Development of Sectoral Action Plan for Nationally Determined Contribution (NDC) implementation road map.

(xiii) Other programs/efforts are as follows:

(a) Energizing Education Program(EEP) for seven universities in Ebonyi, Delta, Anambra, Sokoto, Kano, Benue, and Bauchi states at the cost of N8,553,600,827.75 for 12.5 MW (off grid).

(b) Afforestation programs in Borno, Bauchi, Gombe, Jigawa, Kastina, Adamawa, Kano, Oyo, Plateau, Zamfara, Kaduna, Niger, Oyo, Ondo, Ogun, Abia, and Edo States for N1, 990,331,366.00 covering 1178 Ha.

(c) Renewable Energy Micro Utility (REMU) at Sokoto for N146, 067,806.25 for 60KW (grid connected).

(d) National Trader Moni program under which low income business people are given interest free sum of ten thousand (to inject into their businesses) repayable over a specified period.
(e) Establishment of Climate Change Desk Officers across the states of the federation.
(f) Commencement of the implementation of the NDC with the support of UNDP.
(g) Capacity Building on Measurement, Reporting and Verification (MRV) and GHG Inventory achieved on sector mapping.

(xiv) Nigeria has a number of bilateral and multilateral agreement cooperation with World Bank, Global Environment Fund (GEF) Green Climate Fund (GCF), etc.

In striving to address poverty, it is important to note that significant portions of Nigeria's population and economy are tied to activities that are climate sensitive, such as rain-fed agriculture, livestock rearing, fisheries, and forest products extraction. The heavily populated coastal areas and Northern Sahel zone are particularly vulnerable to climate change. As part of efforts to address this vulnerability, Nigeria, in June 2014, in Equatorial Guinea, together with the Heads of State and Government of the African Union put agriculture on top of Africa's agenda when they adopted the Malabo Declaration on Accelerated Agricultural Growth and Transformation for Shared Prosperity and Improved Livelihoods (Malabo Declaration 2014). In the Declaration, African leaders made several commitments, among them, to end hunger in Africa by 2025.

The discussions above reveal that Nigeria has been making domestic and international efforts at adaptation mechanism and to attain SDGs targets. She has also keyed into bilateral, multinational, and regional initiatives. Although there are policy framework arrangements to manage climate change in Nigeria, in reality, the policies are not harmonized and remain mostly on paper. This is because the legal framework specific to climate change is non-existent and therefore not legally binding. Efforts to manage climate change therefore remain ad hoc and superficial. This is due mainly to the fact that Nigeria has failed to establish a robust climate change law that should incorporate some of the global initiatives and best practices enumerated earlier. This leads us to the need to x-ray the legal issues, challenges, and strategies in the application of climate change adaptation mechanism.

Legal Issues, Challenges, and Strategies in the Application of Climate Change Adaptation Mechanism

There are many legal issues and challenges which hamper the application of climate change adaptation mechanisms in Nigeria. These are examined in this section with a view to proffering strategies to overcome them.

(i) **Lack of legal framework**: In Nigeria, there is the National Adaptation Strategy and Plan of Action on Climate Change for Nigeria (NASPA-CCN) and the

Nigeria Climate Change Policy Response and Strategy (NCCPRS). Nevertheless, despite the well acknowledged position that there is a close nexus between poverty (SDG 1) and climate change adaptation mechanism as earlier articulated in this chapter, Nigeria still lacks the legally binding law on the subject matter. It has been argued at different fora that the non-transmission of the policies into Nigerian laws is in breach of the climate change policy thrust. This is a challenge and one which makes it difficult to assess the effectiveness of such policies. It has however been suggested in some quarters that the birth of Nigeria's Intended Nationally Determined Contribution (INDC) will compel Nigeria to expedite action on establishing relevant laws and become more focused in terms of combating the impact of climate change. Unfortunately, the provisions of the INDC do not appear to have reflected this expectation.

Furthermore, apart from the absence of a law on climate change to galvanize action toward effective policy implementation, there is equally no existing law in Nigeria that provides for climate change management. In fact, until very recently actions taken by the Ministry of Environment Department of Climate Change have been largely ad hoc and divorced from a bigger picture of global fight against climate change. This is due largely to the absence of a climate change law that will regulate the management of every impact of climate change. The challenge is that without a binding climate change law, it will be difficult to achieve effective adaptation mechanism to reduce poverty in line with the SDGs. In terms of strategy to overcome this challenge, therefore, there is the need to establish a legal framework that would comprehensively and legally bind the various policies on adaptation mechanism. This, for example, is the case in Bangladesh which is a good example of the impact a climate change law can have in the struggle to achieve SDG1.

(ii) **Lack of equality:** It has been argued in some quarters that calls to eradicate poverty are meaningless without commensurate and committed policy action to reduce inequality. This is a global menace that must be conquered to stand a good chance of meeting the SDG 1 target. For instance, the income of the poorest 20% of the world's population has only moved from USD $0.94 in 1990 to $1.90 today, but that of the rest of the population has grown from an average USD $12.85 to $18.63 over the same period, thus creating great inequality. This inequality is projected to worsen with an income gap of USD $18.79 by 2030 which is almost double what it was in 1990 (Hope 2018). A recent report by the research organization, Development Initiatives, which looked at the period between 2010 and 2017, found that notwithstanding global commitments to zero poverty as per SDG 1, the attitude of the global actors is at variance with the espoused commitment. Aid donors promised in 1970 to give 0.7% of their GNI to fighting poverty, and yet the average spent up till 2017 is about 0.31%, and this is lower than it was then (0.33%) (Ibid).

In Africa where the population is expanding at a rate of more than 2% per year, a combination of double digit GDP growth and dramatic declines in inequality is a sine qua non for plummeting the level of extreme poverty to below 5% by 2030 and much more than would be expected to bring the level of poverty to 0% by 2030. Corroborating this view, the UN estimates that without significant changes in behavior, more than 7% of the global population may remain in poverty by the year 2030, including about 30% of the populations in Africa and the least developed countries (LDCs). The challenge is for the governments, including the Nigerian government, to bridge this poverty gap.

Since equality is a sine qua non to the attainment of SDG 1, a major strategy should therefore be to ensure that equality is an inclusive part of the fight to eradicate poverty. It is vital that any form of mainstreaming incorporates a poverty focus rather than viewing it as a purely environmental problem. This will ensure that aid initiatives address the vulnerability of the poor to climate changes and equally to align such aids with the priorities of the SDGs. Development partners must continue to support efforts to bridge the poverty gap and to eradicate poverty. In 2010, the International Development Association (IDA) gave almost USD $2 billion more in loans to middle-income countries (MICs) than to the LDCs, whereas in 2017, it gave over USD $2 billion more in loans to LDCs than to MICs (Dodd et al. 2019). A research also found that if actors met their commitments, an extra USD $1.5tn would be available between 2017 and 2030 to help lift the very poorest in the world out of poverty. A number of both bilateral and multilateral donors are driving these trends.

While there is no one-size-fits-all policy prescription that guarantees delivery of a more equal and prosperous society, one overarching message resonates. It is that calls to exterminate poverty are meaningless without concerted and committed policy action for dramatic decrease in inequalities (Zhenmin 2019), in income, in opportunity, in exposure to risk across gender, within countries and between countries. No doubt, well-designed fiscal policies can help smooth the business cycle, provide public goods, enact reforms that would help build resilience, correct market failures which include, for example, raising potential economic growth and its inclusiveness, and directly influencing the income distribution to bridge inequality. Specifically, Nigeria could introduce measures to bring all the poor under coverage of social protection system and anti-poverty strategy of the government through, for example, an expanded Social Safety Net Programmes (SSNP) to address risk and vulnerability as a means of reducing poverty and inequality.

(iii) **Inadequate political willpower:** This entails the lack of sufficient political willpower of government to ensure SDG 1 succeeds, and the inability to elevate SDGs to a very high political level to be able to give it the national attention and priority that climate change issues deserve. Presently, Nigeria does not even have a climate change law to give focus to the fight against climate change. Meanwhile, the political will of the government as good leadership and governance are crucial for implementation of adaptation mechanism. Akin to this is the fact that Nigeria has also not created the favorable climate environment for addressing its long-standing armed conflict or civil unrest, kidnapping, and political instability. This means that decent jobs for most citizens will become increasingly out of reach resulting in weak economic performance. The required strategy to address the problem is for the government to muster enough political courage to elevate critical decision-making for climate change matters to the presidency for needed priority and implementation directives.

(iv) **Centralization of information and services on climate change:** The issue of centralization of information and services at the Ministry of Environment and several other ministries without decentralizing to the door steps of the citizens remains a challenge. This is because adaptation is likely to be successful if people are informed about climate change, how it affects them, and options for doing something about it. Successful climate change interventions are dependent on high-quality accessible information to allow effective decision-making. As the impacts of climate change are difficult to predict accurately, adaptation activities need to be flexible and responsive to new information, and robust enough to withstand a wide range of plausible futures.

There are many examples of use of information to enhance adaptation mechanism. For instance, the use of risk management and coping thresholds is an area of applied adaptation research of growing importance (Part 2: Adaptation). In addition, agricultural climate information is now used to advise farmers about their choice of crops and methods of cultivation, which in turn has provided major benefits in terms of increased yields and preventing food shortages. Similarly, better information and early warning systems for farmers can reduce vulnerability to inter-annual climate variations and enable responses to be proactive rather than reactive. Climate information can generate substantial benefits in other areas as well, including water management, planning and delivery of health services, and improved warning for extreme weather events. The challenge for Nigeria is that in spite of what is provided for in its Nationally Determined Contribution (NDC), Nigeria has failed to decentralize the delivery of public services and take them to the doorsteps of millions of underserved citizens. It has also not actively leveraged the local knowledge of the communities. For example, in Arochukwu, in the Eastern part of Nigeria, farmers can predict a bountiful harvest merely by observing how severe the harmattan is. This highlights the importance of capturing local knowledge, reviewing and assessing its applicability, and its dissemination among other communities and relevant agencies. The strategy to overcome this challenge is to decentralize information. As the poor already have a lot of knowledge about how to cope with climate variability, adaptation activities should take account of this knowledge, where benefits are proven. Incorporating local knowledge into policy actions could help the governments of Nigeria to accommodate specific needs of poor people and ensure that strategies are taken up by local communities. Thus, relevant information must be decentralized from top to the local levels as a matter of deliberate policy.

(v) **Weak human and institutional capacity:** In Nigeria there is weak human and institutional capacity for climate change related issues. This is because climate change is fast altering the landscapes, and the inability of the various government institutions and weak human capacity to respond appropriately has created various risks due to shortages of resources such as land (Orie 2015) and water, with attendant negative secondary impacts, hunger, poverty, sickness, loss of employment, etc. The poor responses have in turn opened the door to conflict. This alteration comes mainly from the arid north; Nigeria's northern Sahel area (the transition zone between the Sahara desert to the north and the grasslands to the south) gets less than 10 inches of rainfall a year already, a full 25% less than 30 years ago; areas of southern Nigeria where recorded volumes of torrential rains increased 20% across various southern states, some of which already see up to 160 inches of rainfall a year, with wet seasons lasting 8–10 months (Odjugo 2005), and along the southern coastline, where sea levels could rise 1.5–3 feet by century's end which is a further increase over the nearly 1-foot rise observed in the last 50 years (Federal Ministry of Environment 2009). (Federal Ministry of Environment, Nigeria and Climate Change: Road to Cop15 (Abuja, Federal Ministry of Environment 2009).) In fact, in 2012 about 26 out of 36 states in Nigeria were submerged in flood water. The situation is getting worse due to inadequate experts and institutions to properly contend with the developments. In Nigeria, there is no conscious effort either toward manpower or institutional development. Most discussions on the impacts of climate change end at the federal level, occasionally at state level but never at the local government level. The implication of this is that there is disconnect between what happens at the federal and state levels and that of the local government. The result is that at the local level people are not motivated to share in the national vision. For example, the locals still engage in illegal sand mining resulting in erosion, land-slides, and other negative impacts on the ecosystem, while at the federal level government is fighting erosion and landslides.

The strategy to checkmate this challenge is to improve human and institutional capacity development at the grassroot level such that a culture of comprehensive risk reduction management would naturally evolve overtime as the locals become stakeholders and team players with the government. Government and private actors also need to ensure that particular adaptive responses do not themselves fuel violence but actively help build peace. Successful adaptation measures will be crosscutting in design and impact, based on inclusive planning and implementation and steer clear of political patronage traps. An example is the system of Payment for Ecological Services (PES) (Fagbohun and Orie 2015).

(vi) **Inadequate needs assessment and financing strategy:** This is a challenge that cuts across the various adaptation mechanisms. It is so because adaptation as a mechanism is not cheap and therefore requires a lot of funding which most of the developing countries like Nigeria do not have. For instance, for the adaptation mechanism to play its role in achieving SDG1, there must be technology transfer from the developed countries to the developing countries, and also data have to be purchased. In the last couple of years, China has been championing transfer of technology through South-South Cooperation. So many member countries of the South-South Cooperation like Bangladesh, Kenya, Ethiopia, South Africa, Angola, etc. have benefitted richly from this collaboration compared to Nigeria. Some of the identified areas are water resources, agriculture, tourism and health, etc. Nigeria should reposition itself to aggressively acquire the much needed technology to adapt to climate change scenarios. No doubt, recent years have seen evidence of China's technical assistance to Nigeria. The argument here is that such assistance would have been much more both in volume and quality had Nigeria been more proactive like some other member countries like Bangladesh and Kenya where CC is elevated to a political level.

In addition, Nigeria is also expected to have done its needs assessment and financing strategy which should be in tandem with its SDG 1. Such an assessment will ensure that aids initiatives address the vulnerability of the poor to climate changes and equally to align such aids with the priorities of achieving SDGs 1. However, the overriding challenge is that without a strong legal framework on climate change, the international community may not be willing to provide substantial aids. Thus, the strategy to overcome this challenge is to do adequate needs assessment and financial strategy, as well as have a CC law in place in order to attract the international community to support Nigeria fully in its adaptation mechanism to attain SDG 1.

Conclusion and Recommendation

In response to achieving the global target of SDG 1 by employing adaptation mechanisms, this chapter examined the legal angle to Nigeria's efforts at combating poverty. This was predicated on World Poverty Clock data that by 2018, Nigeria had the largest extreme poverty population of 86.9 million implying that the people are vulnerable to malnutrition, armed conflict, migration, and other socioeconomic and environmental shocks. It was argued that unfortunately for Nigeria, her adaptation efforts are inadequate to ensure the eradication of poverty by 2030. Consequently, Nigeria needs a comprehensive legal framework which is in line with global best practices to enhance her chances of attaining the objective of SDG 1.

The chapter also established the nexus between climate change adaptation mechanism and poverty reduction, showing a direct positive relationship, such that as climate change-induced poverty occurs, the adaptation measures alleviate the adverse consequences and thus make the attainment of SDG 1 feasible. Conversely, weak adaptation mechanism cannot effectively combat the impacts of climate change and invariably undermines the attainment of SDG 1.

It was shown that although there is a policy framework arrangement to manage climate change in Nigeria, in reality, the policies are not harmonized and remain mostly on paper. This is because the legal framework specific to climate change is non-existent and the various related policies are not legally binding. Efforts to manage climate change remain ad hoc and superficial. Therefore, a way forward was proffered based on some global best practices which Nigeria could leverage in the implementation of adaptation laws.

The legal issues and challenges militating against Nigeria's efforts at effectively employing adaptation mechanisms were discussed. They include lack of legal framework, inequality, inadequate political will, and centralization of information and services. Others are weak human and institutional capacities and lack of needs assessment and financing strategy regarding adaptation mechanism. These challenges were matched with corresponding and relevant strategies which include establishment of a robust climate change law which ought to draw from global best practices and incorporate the extant regulations and policies, curbing inequality through the introduction of social protection system and anti-poverty strategy for the poor to address risk and vulnerability. In addition, the law must provide for climate change and related issues to be elevated to a political level such that government shall muster the necessary political will to domicile climate change related issues at the presidency and give top priority to such issues. Other strategies are decentralization of information down to citizens at the local government level, human and institutional capacity building, and undertaking of needs assessment and financing strategy. The chapter therefore makes the following recommendations.

The chapter thus recommends the establishment of climate change law in Nigeria that incorporates adaptation measures and mainstreaming of climate change issues towards the attainment of SDG 1. This law should be comprehensive enough to give legal backing to the various extant policies. It should cover social protection system and anti-poverty strategy for the poor to address risk and vulnerability; elevation of critical climate change decision-making to the presidency in order to get priority attention and the needed funding; as well as decentralization of information down to citizens at the local government level. Provisions should also be made for periodic training of staff and reform of institutions to effectively undertake climate change and adaptation projects. Significantly, needs assessment and financing strategy should be undertaken to the standards required by development partners and aid givers to attract the necessary levels of foreign funds and maintain transparency and accountability. These recommendations constitute the legal imperatives to properly regulate institutional and policy interventions to impacts of climate change and if implemented will guarantee attainment of SDG 1 by 2030 and thereby enhance poverty reduction in Nigeria.

References

Addis Ababa Action Agenda – the United Nations (2005). Available https://www.un.org/esa/ffd/ffd3/press.../countries-reach-historic-agreement.html

Addis Ababa Action Agenda of the Third International Conference on Financing for Development (Addis Ababa Action Agenda) 2015 (endorsed by the General Assembly in its resolution 69/313 of 27 July 2015). Available https://www.un.org/esa/ffd/ffd3/press%E2%80%A6/countries-reach-historic-agreement.html

Atkinson AB (2016) Monitoring global poverty. Report of the commission on global poverty. World Bank Group, Washington, DC. Available https://openknowledge.worldbank.org/

Ban Ki-Moon United Nations Millennium Development Goals – the United Nations. Available https://www.un.org/millenniumgoals/

Bankoff G (1999) A history of poverty: the politics of natural disasters in the Philippines, 1985–1995. Pac Rev 12(3):8–12

Chen S, Ravallion M (2010) The developing world is poorer than we thought, but no less successful in the fight against poverty. Q J Econ 125(4):1577–1625

CIDA (2001) Canada climate change development fund management framework and business plan. Available cida.gc.ca/cida_ind.nsf/c868c8f732a05e34852565a20067581f/f5036f5adbab9c3385256ad10073ea48?OpenDocument
Climate change is a threat – and an opportunity... (2016) World Bank Group. Available www.worldbank.org/.../climate-change-is-a-threat%2D%2D-and-an-opportunity%2D%2D-for-the-pri...
DFID (2000) Achieving sustainability: poverty elimination and the environment. Available www.albacharia.ma/.../1575Target_Strategy_Paper__Achieving_sustainability_-pover...
DFID, EC, UNDP, World Bank (2002) Linking poverty reduction and environmental management: policy challenges and opportunities, consultation draft of discussion document. DFID, EC, UNDP, World Bank, London/Brussels/New York/Washington. Available documents.worldbank.org/curated/en/347841468766173173/pdf/multi0page.pdf
Dodd et al (2019) Six ways to refocus ODA to end poverty and meet the SDGs. Available devinit.org/post/refocus-oda-end-poverty-meet-sdgs/
Ending Extreme Poverty and Promoting Shared Prosperity (2013) World bank News. Available www.worldbank.org, news, feature, 2013/04/17
European consensus on development | International (2017) European Commission- Fact Sheet. Available ec.europa.eu international-partnerships european-co
Fagbohun OA, Orie EG (2015) Nigeria: *law and policy issues in climate change*, ELRI, monograph series. Available https://kdp.amazon.com/en_US/bookshelf?laguage=en_US
Federal Ministry of Environment (2009) Nigeria and climate change: road to Cop15 Abuja Federal Ministry of Environment
Fiszbein A, Schady N (2009) Conditional cash transfers. World Bank policy research report. Available http://siteresources.worldbank.org/intcct/resources/5757608-1234228266004/prr-cct-web-noembargo.pdf
Global climate change: vital signs of the planet|precipitation... Available https://pmm.nasa.gov/education/websites/global-climate-change-vital-signs-planet
Gomez-Echeverri L (2018) Climate and development: enhancing impact through stronger linkages in the implementation of the Paris agreement and the Sustainable Development Goals (SDGs). Phil Trans R Soc A 376(2119):20160444. https://doi.org/10.1098/rsta.2016.0444
Gurtner B (2010) The financial and economic crisis and developing countries. Available https://journals.openedition.org/poldev/144
Hope A (2018) Launch: investments to end poverty 2018 – development initiatives. Available devinit.org/post/launch-investments-to-end-poverty-2018/
IPCC (1995) Climate change: intergovernmental panel on climate change. Cambridge University Press, Cambridge. Available https://ipcc.ch/ipccreports/tar/wg1/index.php?idp=47
IPCC 2001 – IPCC. Available https://www.ipcc.ch/2001/
IPCC report 2014 fifth assessment report – IPCC. Available https://www.ipcc.ch/assessment-report/ar5/B
Kelman (2017) Linking disaster risk reduction, climate change, and the sustainable development goals. Disaster Prev Manag: Int J 26(3):254–258. https://doi.org/10.1108/DPM-02-2017-0043
Lipper L et al (2014) Climate-smart agriculture for food security. Nat Clim Change 4(12):1068–1072
Malabo declaration on accelerated agricultural growth and transformation for shared prosperity and improved livelihoods (2014). Available https://www.resakss.org/sites/default/files/Malabo%20Declaration%20on%20Agriculture_2014_11%2026-.pdf
McGuigan C et al (2002) Poverty and climate change: assessing impacts in developing countries and the initiatives of the international community. Available https://www.odi.org/sites/odi.org.uk/files/odi-assets/publications-opinion-files/3449.pdf
Medugu NI et al (2011) Drought and desertification management in arid and semi-arid zones of Northern Nigeria'. Emerald Manag Environ Q 22(5):610. Available www.fao.org/3/a-ad710e.pdf. 2003
Messer NM (2003) The role of local institutions and their interaction in disaster risk... Available www.fao.org

National adaptation strategy and plan of action on climate... – CSDevNet csdevnet.org/.../NATIONAL-ADAPTATION-STRATEGY-AND-PLAN-OF-ACTION...
Odjugo PAO (2005) An analysis of rainfall pattern in Nigeria (2005). Global J Environ Sci 4(2):139–145
Olabode PO et al (2015) Determinants of poverty level in Nigeria. J Sustain Dev 8:1
Olawale S (2018) Nigeria poverty statistics and poverty rate in Nigeria – NaijaQuest.Com. Available https://naijaquest.com/nigeria-poverty-statistics/. 5 Feb 2018
Olofin, Adejumo & Sanusi (2015) Determinants of Poverty Level in Nigeria. Journal of Sustainable Development 8(1):235–241. Available https://doi.org/10.5539/jsd.v8n1p235
Olsthoorn et al (1999) The Netherlands, always vulnerable to floods, has a new approach to... Available https://www.pri.org/.../netherlands-always-vulnerable-floods-has-new-approach-water...
Orie EG (2015) Climate change and sustainable development: the Nigerian legal experience. NOUN Curr Issues Nigerian Law 5:68–114
Orie EG (2017) Legal perspective to the mitigation of impact of climate change in Nigeria. Univ Jos Law J 12(1):323–344
Parnell S (2017) Africa's urban risk and resilience. Int J Disaster Risk Reduct 26:1–6
Part 2: Adaptation lessons from past experience – OECD.org. Available https://www.oecd.org/env/cc/4494574.pdf
Portugal Sustainable Development Knowledge Platform (2016). Available https://sustainabledevelopment.un.org/memberstates/portugal
Poverty and Climate Change – odi.org. Available https://www.odi.org/resources/docs/3449.pdf
Poverty|United Nations Educational, Scientific and Cultural.... Available www.unesco.org/new/en/social-and-human-sciences/.../international.../poverty/
Report of the conference of parties on its seventh session held at Marrakesh from 29th October to 10th November 2001: part II – action taken by the conference of parties. Available https://unfccc.int/node/2516
Roser M, Ortiz-Ospina E (2019) Global extreme poverty – our world in data. Available https://ourworldindata.org/extreme-poverty
Sayne A (2011) Climate change adaptation and conflict in Nigeria, United States Institute of Peace Special Report. Available http://www.usip.org/sites/default/files/Climate
Smucker TA et al (2015) Differentiated livelihoods, local institutions, and the adaptation imperative: assessing climate change adaptation policy in Tanzania. Geoforum 59:39–50
Standing Orders on Disaster (2010) – PreventionWeb. Available https://www.preventionweb.net/files/18240_sodapprovedbyndmb.pdf
Sustainable Development, Poverty Eradication and Reducing... – IPCC. Available https://report.ipcc.ch/sr15/pdf/sr15_chapter5.pdf
The global extreme poverty- our world in data. Available https://ourworldindata.org/happiness-and-life-satisfaction/
The Sustainable Development Goals Report 2018|Multimedia Library... Available https://www.un.org/development/.../publications/the-sustainable-development-goals-re...
The World Bank research report (2015) A measured approach to ending poverty and boosting shared prosperity: concepts, data, and the twin goals. World Bank, Washington, DC, pp 37, 38
United Nations Framework Convention on Climate Change. Available http://environment-ecology.com/climate-change.html
Weisser F (2014) Translating the adaptation to climate change paradigm: the politics of a travelling idea in Africa. Geogr J 180(2):111–119
World Bank (1990) World development report, poverty. Oxford University Press, New York
Zhenmin L (2019) Ending poverty is possible, but it means facing up to inequality – within & between countries INTER PRESS SERVICE. Available http://www.ipsnew

7

Climate Change, Biodiversity and Tipping Points in Botswana

Peter Urich, Yinpeng Li and Sennye Masike

Contents

Abstract

Climate adaptation planning requires new ways of thinking and approaching the analysis of risks. Such thinking needs to be systemic in nature and practice/action-oriented while respecting the complexity of the physical and social sciences. Through this chapter on climate tipping points in Botswana, it is proposed

P. Urich (✉) · Y. Li · S. Masike
International Global Change Institute and CLIMsystems Ltd, Hamilton, New Zealand
e-mail: peter@climsystems.com; yinpengli@climsystems.com; sennye@climsystems.com

that a generic and practice-oriented analysis framework be applied with a mathematical foundation including modeling methods based on complex science. The objective is to promote a framework that privileges a worldview to avoid biased and partial explanations of risks. An Institutional-Socio-Earth-Economical-Technical systems (ISEET) approach is based on a systems science philosophy for risk governance analysis, with particular emphasis on tipping points and emergence which are some of the key elements that can support sound adaptation planning. Through the lens of the biodiversity sector in Botswana, the complex interrelationships of ISEET principles are explained. They provide a new, efficient, and practical framework for moving rapidly from theory to action for planning and implementing climate change adaption projects.

Keywords

Tipping points · System dynamics · Climate change · Risk assessment · Biodiversity · Action research

Introduction

Humanity has modified the earth systems in significant ways and has initiated unprecedented Anthropocene risks (Keys et al. 2019; Baer and Singer 2018). Changes include fundamental aspects of the earth system such as all layers of the atmosphere, hydrosphere, and cryosphere through changes in weather patterns, climate, land surfaces, ocean chemistry, and geological structures. Anthropocene risks such as global connectivity either have reached or are approaching tipping points of various earth systems at different temporal and spatial scales (Steffen et al. 2018). Meanwhile, the emergence of new knowledge, technologies, and institutions has led to new approaches for problem-solving. An example of collective action is the Paris Agreement ratified by the United Nations Framework Convention on Climate Change (UNFCCC). However, progress in mitigating climate risk has not been satisfying. Responses are uneven and uncoordinated, and CO_2 emissions continue to increase globally, even with green energy and other technological advancements. New approaches to the analysis of risks need to be articulated systemically and be practice-oriented while respecting the complexity of the physical and social sciences (Sterner et al. 2019; Lucas et al. 2018).

The existing initiatives and frameworks are limited, such as the Sendai Framework for Disaster Risk Reduction, the Paris Agreement, and international programs such as IRGP-IDHP (Integrated Risk Governance Project-International Human Dimension Programme). These limitations transcend the research frameworks as well, such as SES (socio-ecological system), STS (science technology and sustainability, and IAD (institutional analysis and development). The weaknesses of these existing frameworks as analysis and practice tools are being more widely debated and are modified as society seeks a better understanding of systemic risks posed by climate change and large-scale socio-physical disasters (Cole et al. 2014; McCord et al. 2017).

A more generic and practice-oriented analysis framework is explored from a mathematical foundation, including modeling methods based on complex science. A structure of systems thinking applicable in the Anthropocene is proposed whereby analysis privileges a worldview and earth-view that attempt to avoid biased and partial explanations of risks.

Tipping Points and Emergence

There are standard features of systemic risks in different domains. These include the character of agents and emergence phenomena, tipping parameters indicating instability, and more noncommittal empirical observations. Instead, these features can be related as Lucas et al. (2018) describe on fundamental theory for relatively well-understood and straightforward systems in physics and chemistry. A crucial mechanism is the breakdown of macroscopic patterns of whole systems due to feedback reinforcing actions of agents on the microlevel, whereby the role of complexity science forms the basis for unifying the phenomena of systemic risks in widely different domains.

Tipping points sometimes also refer to shifting points. For example, the shift from an unsustainable to a sustainable society requires a radical historical change in the form of a profound transition which could involve a series of connected transitions in many socio-technical systems (e.g., energy, mobility, and food) toward sustainability. People who work from an SES point of view, studying the profound transition and radical change issues, use different expressions (Schot and Kanger 2018; van der Vleuten 2019).

According to Lenton et al. (2008: 1786), the term tipping point refers to a "critical threshold at which a tiny perturbation can qualitatively alter the state or development of a system." The following attributes are identified as tipping points in the Anthropocene:

- Tipping points could impact the whole planetary scale, everything living on Earth.
- Tipping points could impact the transition or the change of regime.
- Interaction crosses boundaries, including administration and nature systems and ISEETS boundaries.
- Both collective and individual social actions operate in multiple sociocultural, technological, governance, biophysical, and knowledge systems which interact with many other systems at the same time and many levels.
- Tàbara et al. (2018) focused on the complexity of attribution and a reductionist approach about systems thinking and the historical drive to an oversimplified explanation of solutions, drivers of tipping points that could improve the likelihood of limiting global warming to either the 1.5 °C or 2 °C target.
- When social and natural scientists collaborate and integrate their studies, new patterns and previously unforeseen relationships that can accentuate understanding have been achieved. Such studies often reveal intricate causes and effects as such works are based on well-defined spatial, temporal, and organizational units,

be they culturally, physically, or politically determined. Another finding in such studies is their non-stationary and as Liu et al. (2009) feedback loops lag in effects from a range of identified causes and their relationship with resilience and thresholds can reveal new insights. It is also recognized that past interrelationships can have spillover effects that can continue to impact on not only a present state of a system but also its future.

- Tipping points are likely to be breached in the future; however, the underlying conditions are challenging to predict, and the accuracy in defining the time and place makes policy and decision-making currently inadequate for either mitigating or possibly avoiding transgressions.
- Thresholds are not constant. Instead, the position of a threshold along a determining variable can change. The consequences of crossing a threshold are context-dependent. The threshold is sometimes known, and the decision-making depends on the effects of crossing it.
- Tipping points or regime shifts are intricately related to the concept of system resilience.

Emergence and Innovation

Emergence plays a central role in theories related to integrative levels and complex systems. Emergence and emergent phenomena are essential concepts in complexity studies (Goldstein 2018). Emergence can be described as either the development or the presence of the existence or formation of common behaviors, whereby the collective actions within a system would not lead to a similar outcome if applied as individual, constituent parts with no recognized interaction. Emergence is also used to describe the properties of a system – what the system does under its relationship with the environment that it would not otherwise complete itself and is coupled to the scope and system boundaries (Ryan 2007). Emergence also refers to the ability of individual components of a system to collaborate, thus leading to rapid and diverse behavioral changes and new features. For the ISEET, more linkages and communication between subsystems, or the building of more relationships among the ISEET subsystems, could lead to new features emerging. These would then reflect the application of the more complex system and possibly more innovative systems thinking. Emergence is typically not reducible to, nor readily predictable from, the properties of individual system components. Therefore, it may appear surprising or unexpected (Halley and Winkler 2008). Emergent phenomena exist in all subsystems, which could provide solutions or options to the existing challenges in all subsystems if the emergence is managed properly (Ceccarelli et al. 2019; Lichtenstein 2014; Roundy et al. 2018).

Emergence could be applied in a conceptual framework. This framework could improve the understanding of scientific and technological progress (Alexander et al. 2012), innovation, and economic growth (Du and O'Connor 2019). Such a holistic innovation system could improve productivity through the diverse knowledge of business resources available. Emergence also could be scored to identify the topics and drivers of innovation (Porter et al. 2019).

The technology innovation process is central to effective adaption to climate change and development challenges. However, models from business and management tend to dominate innovation theory, which sits outside the adaption-development paradigm (Hope et al. 2018). The goal is, however, to support the development of sectoral and technologically detailed and policy-relevant country-driven strategies consistent with the UNFCCC Paris Agreement.

An ISEETS framework can be used to engage stakeholder input and buy-in; design implementation policy packages; reveal necessary technological, financial, and institutional enabling conditions; and support global stock-taking and ratcheting of ambition (Waisman et al. 2019).

Macro-level agreements, such as the UNFCCC Paris Agreement, should be designed to encourage debate on how to tackle climate change through the notion of innovation, applying both technological innovation and marketing issues. Innovation of a technical nature is part of the equation, but it is not the only requirement. It has been suggested by Asayehegn et al. (2017) that an enabling sectoral system of innovation (SSI) be prioritized where some technological innovations contribute to adaptation actions for climate change. Technological impact analysis could be included based on the following approaches: historical sectoral application and improvement on the systems, such as agriculture technology, livestock breeding and feeding technology, ICT technology, carbon sequestration technology, and others where appropriate.

With the emergence of adaptation technology, it can be defined as "the application of technology to reduce the vulnerability or enhance the resilience, of a natural or human system to the risks of climate change" (UNFCCC 2005: 5). Technologies are defined as either "hard" that includes equipment and infrastructure or "soft" such as institutions and management systems (Christiansen et al. 2011). However, some technologies, such as new crop varieties, are not so easy to categorize. Many technologies can be used to address current vulnerabilities to climate and other environmental, economic, and societal concerns and to reduce future exposure to climate change impacts. Some can also be used to address several types of climate change impact in different sectors.

ISEET modeling seems too big to be handled by either a single model or an existing framework. However, when considering specific modeling for a risk governance issue, the data, variables, and parameters could be selected and refined according to their importance and their functionalities. The modeling approaches from different disciplines could be either simplified or reorganized to fit specific purposes. For the subsystem in this chapter, the modeling approaches have been developed for specific contexts, which can be either absorbed or integrated into the ISEET modeling processes.

Modeling systemic or structural change in socio-environmental systems is not new. Tipping point modeling has been carried out by scientists from SES, climate systems, social systems, network analysis, and agent-based modeling disciplines. Based on the ISEET system analysis framework, Fig. 1 depicts an overview of an

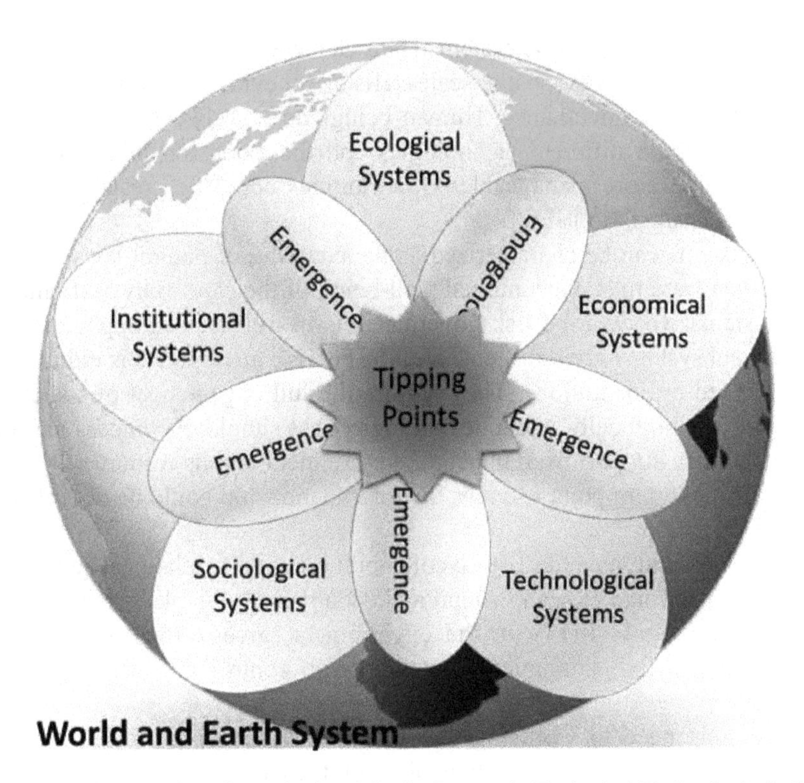

Fig. 1 The framework of Institutional-Socio-Economic-Ecological-Technological Systems (ISEETS) for a climate change tipping point study. (Source: authors)

ISEET structure for tipping points and the emergence of, and linkages between, five subsystems.

Tipping points and regime shifts are being studied within various disciplines applying a range of modeling approaches and analysis frameworks. One of the first obstacles in any study of systemic change is the terminology. Environmental and social science disciplines have engaged in relevant ideas of regime shift, structural change, non-marginal change, and transition theory, and each claims ownership (Polhill et al. 2016).

ISEET Analysis Framework

Institutional-Socio-Earth-Economical-Technical (ISEET) systems

ISEET describes five intrinsically interlinked systems formed around the Anthropocene. The method has at its foundation systems science concepts and mathematical methodologies for risk governance analysis, with emphasis on tipping points and emergence, which are the key characteristics that need to be analyzed for risk governance.

ISEET represents a dynamic system. Their elements, functions, and relationships change with time and across spatial scales. Historical evidence and future prediction are essential for risk governance. Human beings do not always passively react to risks. With new and different technologies and the extension of knowledge, the risk may become more manageable. Opportunities could be created through the emergence mechanism in ISEET.

ISEET systems can be characterized as the coupling of natural laws with world rules. The goals are to support mutual well-being of the earth (physical) and world (social) systems. To describe risk governance issues in the Anthropocene, ISEET framework subsystems are indispensable. ISEET risk governance is either realized or implemented by institutions. They require the full engagement of societies and apply certain economically viable technologies that should encompass sustainable ecosystem service support from the earth environment and its resources.

From the subsystem point of view, the interrelationship could be described as:

- An institution that must have close collaboration with its relational society, using environmentally and ethically sound and economically viable technologies.
- Societies that could live with the environment, given effective and efficient institutions, and be economically healthy, with controllable risks on the earth system.
- Societies that need to work together to advance and apply technologies with economic endeavors, to live sustainably with the earth system.
- The sustainable development of economic systems that could be achieved by a well-designed and well-operated institution, which include social capital and earth resources.
- Technology that may need to be advanced and applied and that does not harm the environment and is ethically healthy. In this way, societies are more likely to promote institutions, with the input from their constituent economic systems.

The Working Definition of a Subsystem

- An institution in ISEET implies a body that operates within regulations, laws, policies, and conventions. Such institutions can represent cross-sectoral entities established when either developing or implementing a related risk governance framework. For example, emergency management laws and emergency management ministries are all part of a larger institutional system.
- Social systems in ISEET refer to all the elements, functions, and relationships in societies, including the population and its age and gender composition, culture, religion, educational level, and connectivity. Complexity theory offers the toolkit needed for this paradigm shift in social theory (Walby 2007).
- Earth system refers to the biophysical existence of the Earth planet, including the living support environment, from top air to deep earth.
- An economic system, as defined by Gregory and Stuart (2013: 30), "is a system of production, resource allocation and distribution of goods and services within

a society or a given geographic area. It includes the combination of the various institutions, agencies, entities, decision-making processes and patterns of consumption that comprise the economic structure of a given community."

- Technological systems are sets of interconnected components that transform, store, transport, or control materials, energy, and information for specific purposes. Machines, software, and the hardware they run on are considered part of a technological system. Similarly, how humanity organizes itself to apply such technology are broader arrangements based on the organization's structures to exploit technology and techniques developed to optimize their application.

Case Study: Botswana's Biodiversity Sector

Time is of the essence for engaging in the intersection between climate change and the biodiversity extinction crisis. The Global Deal for Nature (GDN) is one opportunity as it is science-driven with the goal of saving the diversity and current relative abundance of life on the planet. The linkage of the GDN with the Paris Climate Agreement might help humanity avoid catastrophic climate change while conserving species and their increasingly recognized values, including ecosystem services.

Compelling recent findings add additional urgency to the issue as less than half of the globe's terrestrial realm is intact. The application of global climate points toward a tipping point. Habit conversions continuing as they were historically while greenhouse gas emissions maintaining its current trajectories may exceed humanity's chances of limiting global warming to the 1.5 °C target. Over the next 10 years, currently expanding conversion and poaching rates need to be slowed down considerably to avoid "points of no return" for some floral and faunal species (Fig. 2).

If global mean temperatures are permitted to rise above 1.5 °C, it is widely believed that fundamental aspects of ecosystems, both large and small, could unravel. Continued unsustainable use of the natural environment threatens our global health as witnessed by the rising risk of global pandemics, while mass migration owing to the lack of access to resources such as clean water and productive and uncontaminated land become more widespread. Global climate change and its increase in extreme events could accelerate the degradation of land and societies. For example, climate change-induced sea level rise and extreme still high-water events, which inundate coastal zones and droughts, may displace at least 100 million people by 2050. Most of those people currently live in the southern hemisphere (Dinerstein et al. 2019).

Botswana, as part of Southern Africa, is home to an appreciable portion of global biodiversity, and many of its ecosystems retain relatively intact species assemblages across all trophic levels. The region possesses an established network of protected areas that contribute both to conservation targets and to nature-based tourism. Pressure on biodiversity can result from regional and highly localized developments concerning extractive resource use. Anthropogenic climate changes are more widely accepted as a profound driver of such impacts for Africa's biodiversity, including

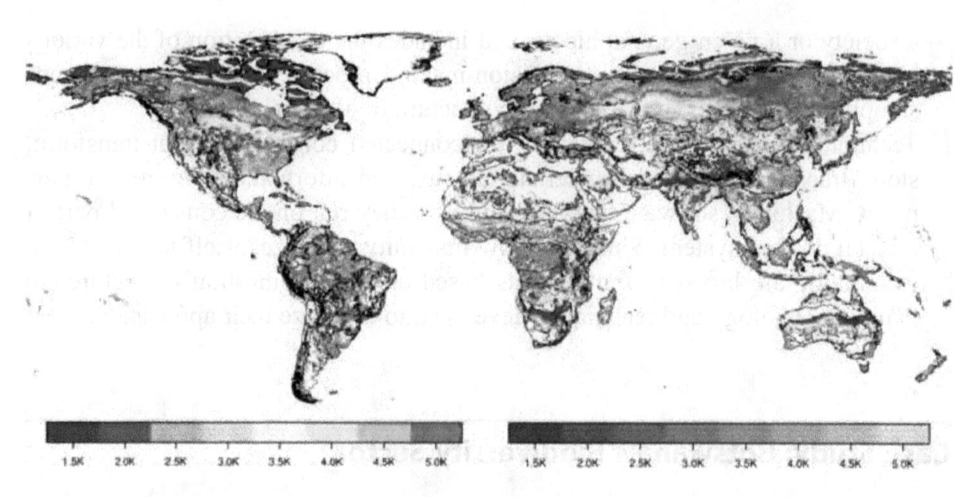

Fig. 2 Thresholds of temperature anomaly that could lead to significant local changes in land-based ecosystems. Colored areas (left legend) represent regions with severe transformation; gray areas (right legend) are likely to experience moderate transformation. Dark red transformation at 1.5 °C and light red at 2.0 °C. (Courtesy of Gerten et al. 2013)

Botswana, and it is increasingly likely to be harmful from both ecological and economic perspectives.

Worldwide, the United Nations is committed to mainstreaming biodiversity planning in a wide range of policies and programs and with the inclusion of climate change. In a country such as Botswana, mainstreaming biodiversity and/or wildlife management at the local, regional, and country level is critical to its economy and its place as a critical biodiversity REDD hotspot for several primary faunal species.

Botswana, in terms of biodiversity, is a country of contrasts. The diversity ranges from the wetlands of the north, dependent on water arriving season from neighboring Angola, to the broad and arid Kalahari Desert in the center and southwest. Each area is part of a systematic protected area system with the Okavango Delta representing the world's largest inland delta which is also a Ramsar site. At the same time, Chobe National Park has many varieties and populations of game and a considerable density of elephants. The country has innovated the first formally declared trans-boundary park in Africa, the Kgalagadi Transfrontier Park. There is also the Central Kalahari Game Reserve and the distinctive prehistoric lake that now consists of salt pans called the Makgadikgadi and Nxai Pans National Park system. This area has important habitats for migratory birds.

Botswana is developing a National Climate Change Policy, Strategy and Action Plan (NCCPSAP) with the framework for such being only recently devised. The policy will be implemented through the Ministry of Environment, Wildlife and Tourism in cooperation with the United Nations Development Programme. Among other objectives, the NCCPSAP aims to develop and implement appropriate adaptation strategies and actions that will lower the vulnerability of Botswana and various sectors of the economy to the impacts of climate change.

Biodiversity and Tipping Points

Biodiversity in the context of tipping points is a complex area for investigation. Alternate stable states are associated with abrupt shifts in ecosystems, tipping points, and hysteresis, all of which challenge traditional approaches to ecosystem management (Oliver et al. 2015). Ecosystems often maintain their stability through internal feedback mechanisms. Environmental perturbations (natural and enhanced by climate change) can change the frequency and magnitude of regime shifts leading to fundamental changes in the assemblages of species providing functions.

Systems can be more susceptible to environmental randomness/irregularity and perturbations/fluctuation close to these critical tipping points and can lead to sudden changes and foster a new equilibrium. Such evolved alternative stable states might be unsupportive in terms of ecosystem functions with a return to a previous state only possible through substantial and costly management interventions (hysteresis). Therefore, the recovery capacity of ecosystem function can be compromised. Alternative conditions have been documented in a wide variety of ecosystems from local to global scales. However, how stable and persistent these will be in the future, under rapid changes in climate, remain uncertain.

It is exceedingly difficult to understand how complex ecosystems, for example, the Okavango Delta, will behave as they either approach or surpass tipping points (Fig. 3). Exceedingly small changes in one or more conditions can lead to a cascade of other changes resulting in a large shift in the state of the system. Sometimes this process can be played out very slowly and therefore less perceptibly by society, and at other times, extreme events lead to radical shifts that exceed the capability of systems to slowly recover, and hence it is forced to find a new stasis; this is a natural process with volcanic eruptions, earthquakes, and flood events often leading to abrupt changes. However, climate change and its speed of onset, which varies from location to location, and the uncertainty in future projections can lead to management paralysis. The slowing down of a potentially catastrophic collapse of individual and highly relevant parts of a system is therefore preferred to prolong the sustainability of the large system.

Ecosystem management depends on monitoring and maintaining resilience because the loss of resilience renders ecosystems more vulnerable to undesirable shifts. Several works summarized by Dai et al. (2012) suggest that a set of generic indicators may aid in the sustainable management of fragile ecosystems. Signals of critical slowing down based on time series demand observations over a long span. Compiling such data is often tricky; therefore, if other indicators based on the spatial structure can be identified, they could be complementary to the early warning signals.

Tipping points are often driven by either complex feedback mechanisms or interactions between multiple drivers. Some of these triggers or drivers are found to be new and thus are not well represented in models currently in use (Leadley et al. 2010). An example is a relationship between dying back in Amazon forests and deforestation and climate change processes that resulted in an underestimation of impacts in earlier global biodiversity assessments. This situation may pertain in the

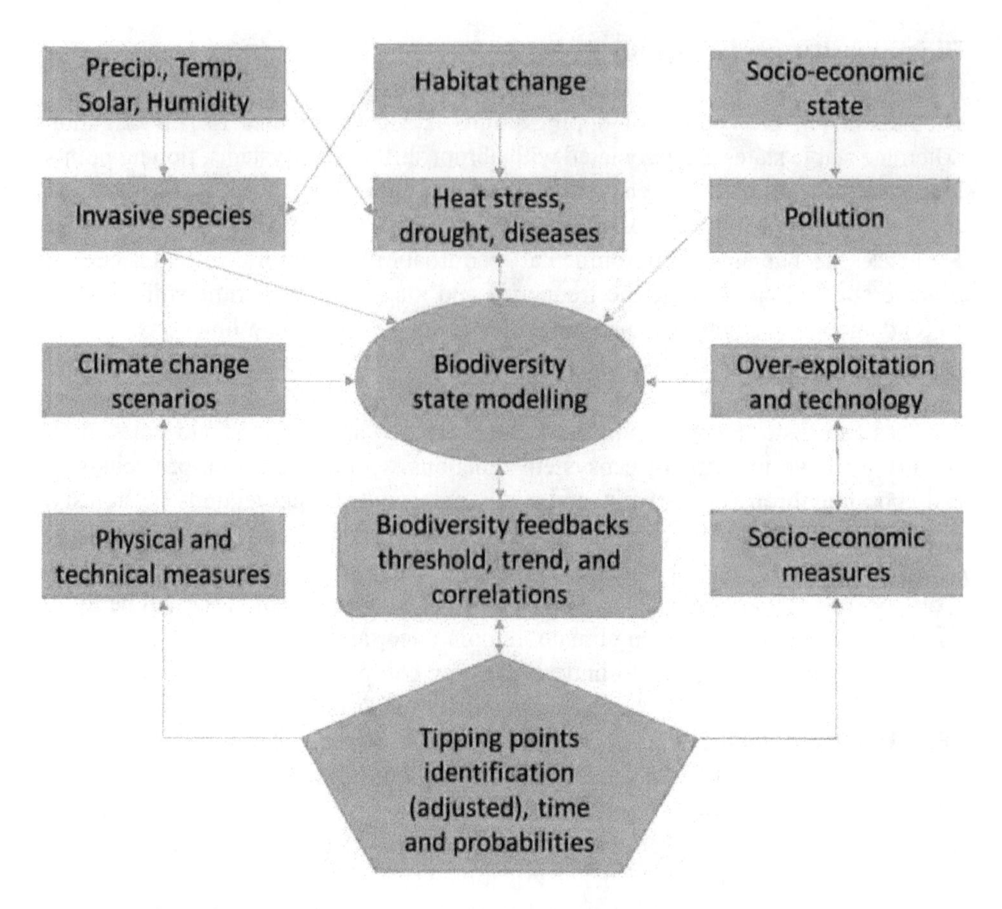

Fig. 3 Tipping point analysis framework for biodiversity in Botswana. (Source: authors)

case of the Okavango Delta where the sustained integrity of the Miombo woodlands on neighboring country of Angola leaves the well-established ecotourism focal areas such as the Okavango swamps in Botswana critically dependent on the sustained flow of sediment- and nutrient-free water from those upland parts of the wider river basin (Leadley et al. 2010).

Ecosystem service degradation can be linked with species extinctions, eroding species abundance, or as shown in this chapter potential shifts in biomes and associated species distributions. However, conservation of biodiversity and provision of some types of ecosystem services can conflict. The following tables list the already identified changes (from 2008 to 2019) in red category lists for plants and animals in Botswana in a limited African context (Tables 1, 2, and 3).

Midgley and Thuiller (2011) suggest significant impacts from unrestrained climate change for the southern part of the African region. However, overestimates of the speed of change and extent of those impacts may be the case owing to underlying assumptions of bioclimatic modeling. The analysis of a diverse range of studies does, however, support the rationale for a high level of concern as there is a signal across the available research that unmitigated changes in climate threaten a

Table 1 Red List Category summary country totals (plants) by the number of extinct, threatened, and other species of plants in each Red List Category in each country (IUCN 2008, 2019)

Sub-Saharan Africa	EX	EW	Subtotal	CR	EN	VU	Subtotal	LR/cd	NT	DD	LC	Total
Angola 2008	0	0	0	0	2	24	26	0	6	1	6	39
Botswana 2008	**0**	**0**	**0**	**0**	**0**	**0**	**0**	**0**	**3**	**0**	**3**	**6**
Angola 2019	0	0	0	0	4	32	36	14	0	38	859	947
Botswana 2019	**0**	**0**	**0**	**0**	**1**	**2**	**3**	**2**	**0**	**8**	**398**	**411**

IUCN Red List Categories: *EX* extinct, *EW* extinct in the wild, *CR* critically endangered, *EN* endangered, *VU* vulnerable, *LR/cd* lower risk/conservation-dependent, *NT* near threatened (includes *LR/nt* lower risk/near threatened), *DD* data-deficient, *LC* least concern (includes *LR/lc* lower risk/least concern)

Table 2 Red List Category summary country totals by numbers of threatened species (critically endangered, endangered, and vulnerable categories only) in each major taxonomic group by country (IUCN 2008, 2019)

Sub-Saharan Africa	Mammals	Birds	Reptiles*	Amphibians	Fishes*	Mollusks*	Other inverts*	Plants*	Fungi and protists*	Total*
Angola 2008	14	18	4	0	22	4	1	26		89
Botswana 2008	**6**	**7**	**0**	**0**	**2**	**0**	**0**	**0**		**15**
Angola 2019	19	32	7	0	53	7	4	36	0	158
Botswana 2019	**11**	**16**	**1**	**0**	**2**	**0**	**0**	**3**	**0**	**33**

*Reptiles, fishes, molluscs, other invertebrates, plants, fungi & protists: please note that for these groups, there are still many species that have not yet been assessed for the IUCN Red List and therefore their status is not known (i.e., these groups have not yet been completely assessed). Therefore the figures presented below for these groups should be interpreted as the number of species known to be threatened within those species that have been assessed to date, and not as the overall total number of threatened species for each group

Table 3 Red List Category summary country totals (animals) by the number of extinct, threatened, and other species of animals in each Red List Category in each country (IUCN 2008, 2019)

Sub-Saharan Africa	EX	EW	Subtotal	CR	EN	VU	Subtotal	LR/cd	NT	DD	LC	Total
Angola 2008	0	0	0	9	19	35	63	0	50	114	1,384	1,611
Botswana 2008	**0**	**0**	**0**	**1**	**1**	**13**	**15**	**0**	**21**	**12**	**847**	**895**
Angola 2019	0	0	0	12	33	77	122	60	0	233	2,493	2,908
Botswana 2019	**0**	**0**	**0**	**4**	**7**	**19**	**30**	**29**	**0**	**10**	**1,010**	**1,079**

IUCN Red List Categories: *EX* extinct, *EW* extinct in the wild, *CR* critically endangered, *EN* endangered, *VU* vulnerable, *LR/cd* lower risk/conservation-dependent, *NT* near threatened (includes *LR/nt* lower risk/near threatened), *DD* data-deficient, *LC* least concern (includes *LR/lc* lower risk/least concern) considerable portion of southern African biodiversity. It is the underlying shifts in ecosystem structures, for example, increases and decreases in woody plant cover, that have a secondary impact on faunal diversity that is likely to alter the dominant savanna vegetation type of the region. Midgley and Thuiller (2011) pointed to the winter rainfall areas of the broader region that could suffer the most significant biodiversity loss. The trends identified in Botswana are echoed in other biomes. It is increasingly recognized that rates of disturbance vary with time and can depend on long-term climate trends, the influence of anthropogenic land-use practices (e.g., fire), wildlife population cycles, and other factors such as presence or introduction of invasive species (Wilson et al. 2019). As noted by Wilson et al. (2019), assessment of regional patterns and trends is needed, hence our approach that placed Botswana in the context of Southern Africa. Specifically, some crucial areas are mostly outside Botswana, but they have relevant spillover effects, especially for the Okavango Delta.

Specifically, when mammals in the region are differentiated by size and dietary requirements, some more telling climate risks emerge. Correlations are significant for annual temperature but only for large mammals, where 60–67% of the variability in species richness of large mammals is impacted versus <20% for small mammals (Andrews and O'Brien 2000). Small mammals are, however, strongly correlated with other either climatic or vegetation parameters. Plant richness, thermal seasonality, and frugivorous and insectivorous mammal richness are found to be correlated with thermal seasonality and minimum monthly PET (potential evapotranspiration). It is also found that arboreal and aerial species richness is associated with plant richness, thermal seasonality, and minimum monthly PET.

It is clear from Andrews and O'Brien's work (2000) that different classes of mammals respond to climatic and environmental factors in important ways. Earlier studies they contend did not identify these discrepancies, as the distinction between various sizes and guilds of mammal was not a common factor of analysis. With climate change, there will be issues across the diversity of mammalian species that should impact on their conservation in the future. This also points to the complexity of communities and the understanding of the potential importance of indicator species, whereby either a mammal's or other organisms' presence, absence, or abundance reflect on the environmental condition of either the ecosystem or biome. Such species can serve as critical indicators of tipping points in either an ecosystem or biome and act as a proxy for the health of that environment in the face of a changing climate.

Baseline Biodiversity in Botswana

Wide-ranging large carnivores are common in Botswana. They often range beyond the boundaries of protected areas into human-dominated areas. The mapping out on a country-wide scale and identifying areas with potentially high levels of threats to extensive carnivore survival is required when formulating national conservation action plans whether considering climate change or not. For this chapter, NPP (net primary production) in a historical context (and later in the context of climate change) has been linked with species biodiversity and richness.

This was done as the notion of NPP was linked with the country's large carnivore guild as part of a recent mapping project. In that project, Winterbach et al. (2014) identified and mapped areas consisting of leopard (*Panthera pardus*), lion (*Panthera leo*), cheetah (*Acinonyx jubatus*), brown hyena (*Hyaena brunnea*), spotted hyena (*Crocuta crocuta*), and African wild dog (*Lycaon pictus*) (Fig. 6). They discovered through this mapping project that habitat suitability for large carnivores depended primarily on prey availability but with secondary relationships with interspecific competition, plus conflict with humans. Winterbach et al. (2014) found that prey availability was a critical natural determinant. Wild ungulate species were preyed upon by the six large carnivores, and this helped to identify different management zones for large carnivore populations. The relationship with large ungulates and NPP

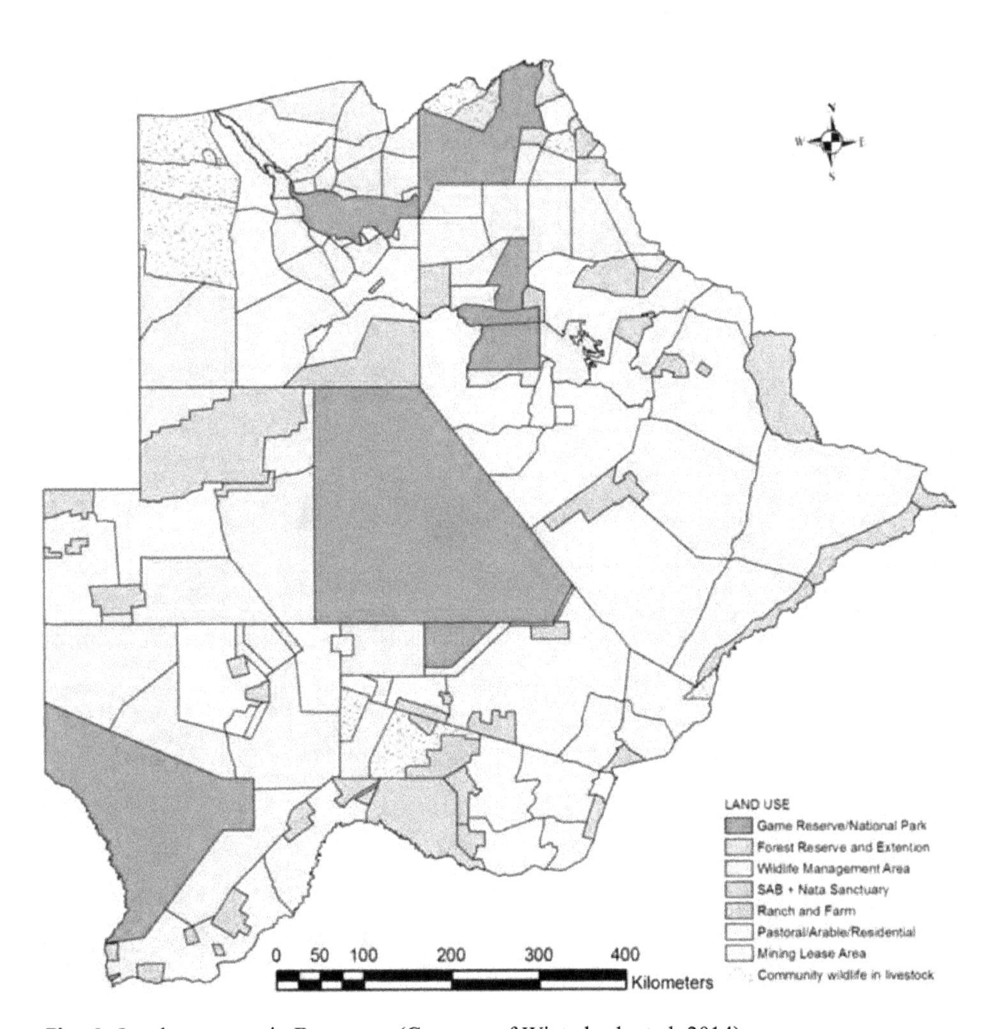

Fig. 4 Land-use zones in Botswana. (Courtesy of Winterbach et al. 2014)

in the country is strong. Therefore, small prey biomass was more evenly spread across large parts of the country, while high to medium biomass of large prey was primarily confined to conservation zones (Figs. 4 and 5).

As Botswana is central, both geographically and in terms of species biodiversity of the larger Southern African sub-Congo basin, its ecological system was deemed prudent to map changes in net primary production and species richness regionally. Changes historically and in the future in these factors will mean transboundary issues will ensue. It is evident in Fig. 6 that mammal species richness is linked with regional phenomena such as NPP, annual precipitation, and mean changes in temperature. The future will be different; hence, some biome shifts from areas currently outside the borders of Botswana could encroach into Botswana over time and as temperatures rise. Clearly, this could lead to critical transboundary implications for management of the subsequent changes in floral and faunal biodiversity.

Future changes in the three zones of species richness in Botswana (Fig. 7) across three key indicator groups, mammals, birds, and amphibians, show clear patterns from the southwest (low richness across all three groups nationally) to moderate richness in the east and high richness in the north. Importantly, these three zones are not national boundary bound but extend to neighboring countries as shown in Fig. 7.

Fig. 5 Entrance to Moremi Reserve. (Source: authors)

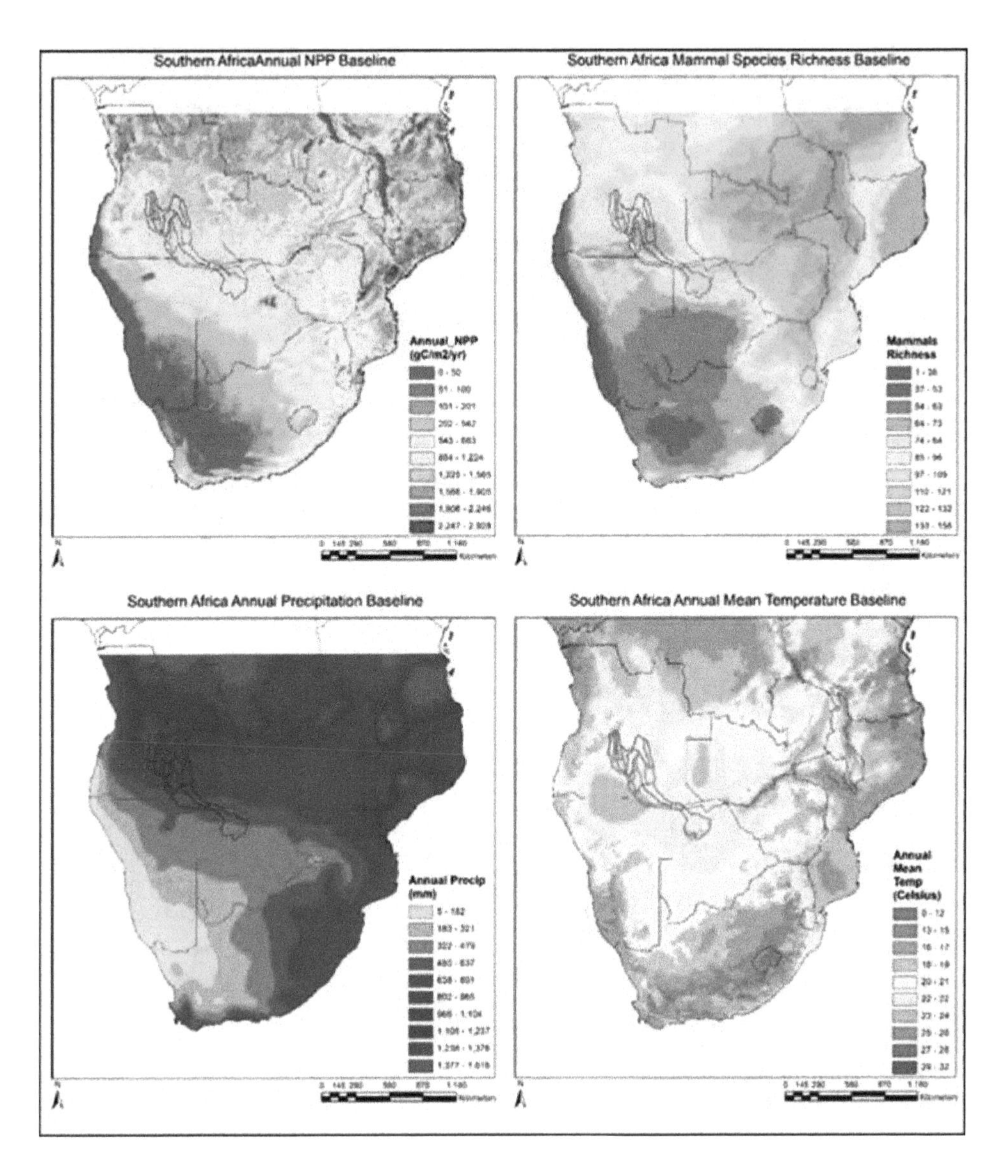

Fig. 6 Baseline NPP and the relationship between high values and mammal-specific richness, annual precipitation, and annual mean temperature. The Okavango River Basin that supports the Okavango Delta in Botswana is included as it is a key transboundary basin for Botswana species diversity and concomitant tourism and economic activities. (Source: authors)

Biodiversity, Ecosystem Services, and Tourism

Wildlife tourism in Botswana has provided strong economic incentives for conservation. With an abundance of wildlife and either the presence or absence of high-profile species, some zones are more suited to wildlife tourism. Winterbach et al. (2014) developed a set of parameters for assessing wildlife abundance and diversity.

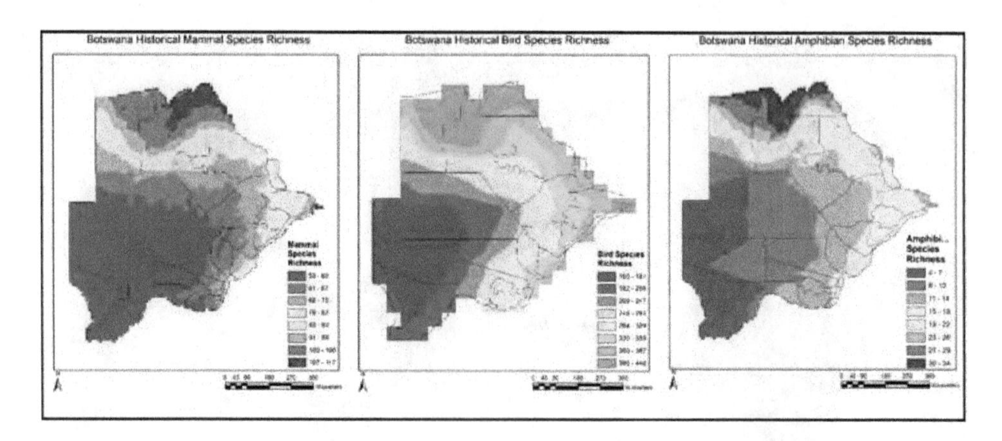

Fig. 7 Historical mammal. Bird and amphibian species richness highlight three key zones, the southwest (low), east (moderate), and north (high), and these three regions extend to neighboring countries. (Source: authors)

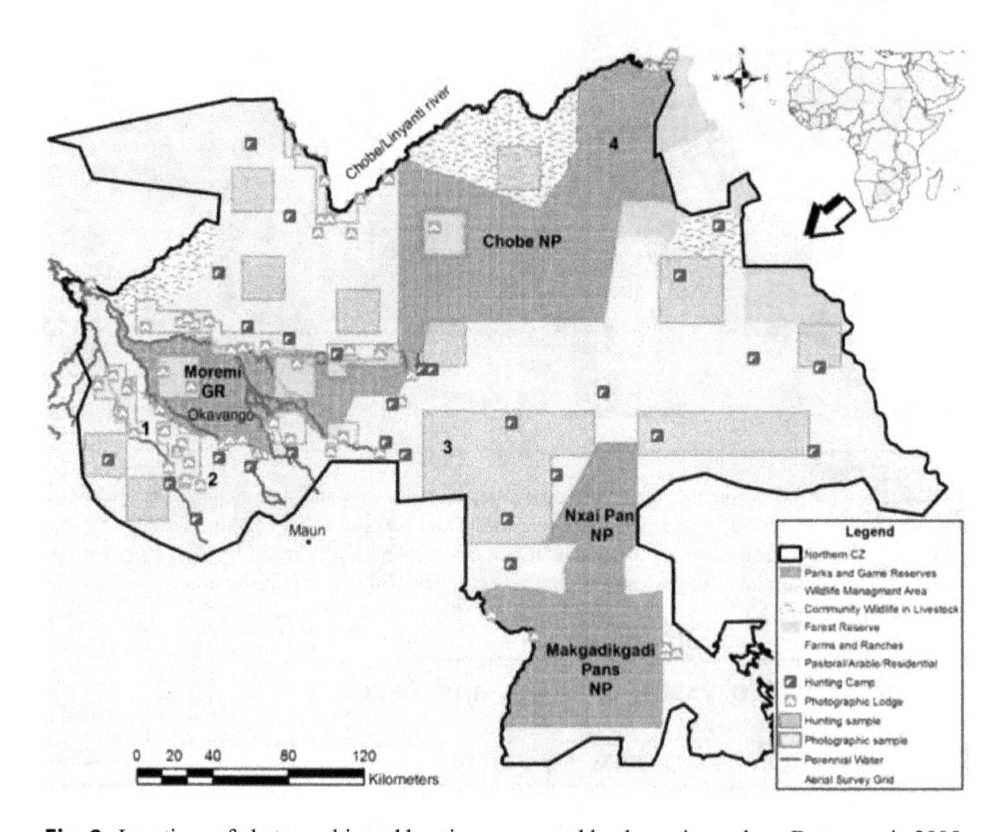

Fig. 8 Locations of photographic and hunting camps and land uses in northern Botswana in 2005. (Courtesy of Winterbach et al. 2015)

Their goal was to evaluate tourism potential in Botswana's Northern Conservation Zone. They also quantified and compared tourism experiences in areas with high and low tourism potential. Wildlife biomass is used as an indicator of NPP, and actual tourism experiences were included in their study to validate their mapping exercise.

Winterbach et al. (2014) found that areas used for high-paying/low-volume tourism had significantly higher mean wildlife biomass and diversity (Fig. 8). Only 22% of the Northern Conservation Zone against their framework had intermediate to high tourism potential. High tourism potential areas afforded tourists better wildlife sighting based on aerial survey data than low wildlife sights based on ground surveys. They also found that the economic viability of low-paying tourism may not even be met in zones with intermediate to high tourism potential. Much of the Northern Conservation Zone was found to have low tourism potential; however, this does not equate with a low conservation value.

Climate Change Impacts on the Biodiversity in Botswana

Africa is expected to be particularly severely impacted by climate change (IPBES 2018). Biggs et al. (2008) assessed the impacts on terrestrial biodiversity using the Biodiversity Intactness Index (BII). This index defines an average change in population size against a pre-modern state, across all terrestrial species of plants and vertebrates. In the next 100 years, they projected a decline in average population sizes of taxa that exceed by two to three times the decline that occurred since circa 1700. A reduction in this modeled declines of biodiversity loss in southern Africa posed considerable challenges. Better alignment is needed for integrated biodiversity conservation and development priorities in the region. Furthermore, what is thought to be required are context-sensitive conservation targets that account for the development of imperatives in different parts of the region.

The predominant climate signal from the climate models for southern African is for a hotter and drier future. This is having, and will likely continue to pose, serious consequences and challenges for policy development and management of environmental and related social change for the people of the region. Historical climate records, as indicated, fit reasonably well with future projections for the region, with the important caveat that temperature rises are trending faster than the global mean, especially in continental interiors such as Botswana. Rainfall, however, apparently has not changed significantly, as yet although the combination of higher temperatures and the potential for changes in the variability of rainfall (e.g., more intense) and also changes in seasonality has already shown changes in the trends in vegetation trajectories in the major biomes of southern Africa. Contrary to early projections for the Succulent Karoo biome, biomass and cover increased over time. This was mainly deemed in response to changes in land-use practices. MacPherson et al. (2019) also found that the fire-adapted fynbos biome either remained stable or increased over time with some expansion of forest species.

Meanwhile, and in the same study, it was found that the shrub-dominated Nama-Karoo biome increased in grass cover. Significantly, and in contrast with earlier model predictions, the grassland biome has expanded westward into former Nama-Karoo biome sites. Regionally, the predominant savanna biome has experienced a relatively rapid increase in woody plants at rates initially unexpected by climate/vegetation models (MacPherson et al. 2019). The availability of new model parameters for biomes and net primary production also provide a useful context for evaluating future changes and for considering tipping points.

Climate change and adaptation have also become a significant issue in contemporary tourism development and policy discussions. This is particularly the case in Botswana and the broader geography of Southern Africa where wildlife in its natural environment dominate the sector. A study on perceptions of climate change (Saarinen et al. 2012) and adaptation strategies of tourism operators in Kgalagadi South District, of southwest Botswana, which also looked at their adaptation strategies, found a general awareness, but the research was conducted nearly 10 years ago. At that time, most operators did envisage challenges to future business growth and Botswana's tourism competitiveness. The perception is that climate change did not at that time have any impacts and that few adaptation strategies were in place. Anecdotal evidence from the consultant's time spent in the Okavango Delta (2019) with a tourism operator of considerable experience pointed to heightened concern regarding climate change. However, problems with earthquake impacts and climate change were signaled as underlying factors in reduced water flow into various parts of the Delta.

One Example: The Okavango Delta

The Government of Botswana has adopted a policy of economic diversification, which is reflected in the National Development Plan 8. There is a strong emphasis on the sustainable use of renewable resources such as veld products and wildlife. Tourism has been identified as a potential "engine for growth" (Jones 2017) and that tourism is based on wildlife and nature experiences. There is an increasing emphasis on the conservation and sustainable use of these resources with a strong geographic focus on the Okavango Delta.

There are significant variations in the size of the actual wetland seasonally as well as from year to year, depending on rainfall intensities in the watershed and other factors (McCarthy 2003). About half of the wetland is permanently inundated, whereas the other half is only seasonally flooded (Anderson et al. 2003); the two areas are referred to as the permanent swamp and the seasonal swamps, respectively.

The Okavango Delta is important ecologically as well as economically. The remoteness, spectacular landscape, and richness in wildlife make the Okavango Delta a magnet for tourists, and tourism has become the second most important sector of the Botswana economy.

Climate change and development-induced land-use change are explicit threats to sustaining biodiversity (Newbold 2018). Modeling and data limitations have hampered a better understanding of the underlying systemic risks. Also problematic are the quite different scales at which land-use and climate processes operate. Newbold

Fig. 9 Differentiation in water distribution within the Delta. (Source: authors)

(2018) predicted that climate change effects will become a significant pressure on biodiversity and could exceed the effects of land-use change by 2070. Both pressures were predicted to lead to an average cumulative loss of 37.9% of species from vertebrate communities under a "business as usual" scenario with an uncertainty range from 15.7% to 54.2%. The biomes to face the most significant pressures were tropical grasslands and savannahs (Fig. 9).

Tipping Points for Climate Change for the Biodiversity Sector

Results and Recommendations

Climate change is already well advanced in Botswana and has already created new challenges for biodiversity conservation. Evidence shows that two key elements that are driven by changes in climate are in a national-scale tipping point analysis: shifting biomes and changes in species composition that relates to net primary production of specific environments. Owing to the complexity of biodiversity and variability of species richness across the country, it was only possible to define broad relationships in terms of potential tipping points.

As mentioned, the dominant relationship between net primary production and biome shifts with the potential to support stable biodiversity mixes is complicated. Natural succession of species must be disentangled from a complex web of confounding factors including a plethora of human factors as well as a relatively rapid change in climate regimes. There appears to be emerging a maturing of the science related to shifts on biomes with changes in climate and signals are thus becoming more transparent that climate has a more dominant role to play in the transition of landscapes and hence biodiversity.

The mapping of mammalian, bird, and amphibian species in Botswana (Fig. 7) with modeled climate-induced changes in net primary production provides insights into possible future changes as temperatures continue to rise regionally. Of note, the temperature profiles of 1.5 °C, 2.0 °C, and 3.0 °C are presented in the introduction and are expected to be reached under current business as usual scenario (84th percentile) (RCP 8.5) in roughly 2028, about 2036, and around 2050, respectively. There will, of course, be lag times in the response regarding average net primary production in the regions and subsequent potential shifts in biodiversity, barring any management interventions and other human dimensions of change that could ensue over the next 30 years.

The trend in net primary production, which is a driving factor in species richness, is for expansion from southwest to northeast of the lower productivity zone. Concurrently, across all three future temperature profiles, the higher NPP areas of the east and north also become progressively less productive. The same trends are then seen in Fig. 7 where the mammalian species richness diminishes nationwide, i.e., there are no parts of the country that with a temperature change should expect to see an expansion in species richness under natural conditions and changes in the net primary production. Again, these projections are made barring human dimensions of change and its management interventions. The same situation pertains to bird and amphibian species richness with a general overall reduction nationwide but with a general trend of a decline in net primary production from the southwest to northeast.

The change in NPP and species richness while following these trends also presents the opportunity to look contemporarily at areas with current NPP and species richness and consider how economic development strategies may intersect when the conditions found in one area expand into others and transition other biomes and communities. Therefore, there are current activities in the zones that will develop that may present insights for the future as these zones shift into new geographies. Given the relative homogeneity of the Botswanan topography, such transitions could expect to vary relatively little from current forms as they expand. Importantly, there is a change in NPP at the 3.0 °C tipping point that is different. In all the species richness projections (Fig. 7) at this degree of warming, a new low-end species richness category could exist in Botswana that is currently not represented. This is presented in the species richness maps as the 42–48 category for mammalian and bird species richness and the 2–4 range for amphibian species richness. Areas with these ranges of species richness could be found to the west and southwest of Botswana currently, but analysis would need to be done to see if these are, in fact, presently located in the broader region or this new low-end category is currently unknown in the area (Figs. 10, 11, 12, and 13).

Related to the shifts in NPP and species richness described above, the ranges and ecological dynamics of the country are changing, and current reserves will possibly struggle to support all species they were designated to sustain. Climate is, however, just one of the myriads of ISEET issues that impinge on the sustainability of Botswana's biodiversity. Action to mitigate the deterioration in biodiversity varies by spatial scale and the actors required. It is generally viewed that adaptation will require more and better regional institutional coordination, which broadens spatial and temporal perspectives with climate chance scenarios being considered as integral to the planning of actions. Local communities are critical to attaining conservation goals, as there are multiple threats to success. Regional planning, site-scale management, and the continued assessment and modification of conservation plans

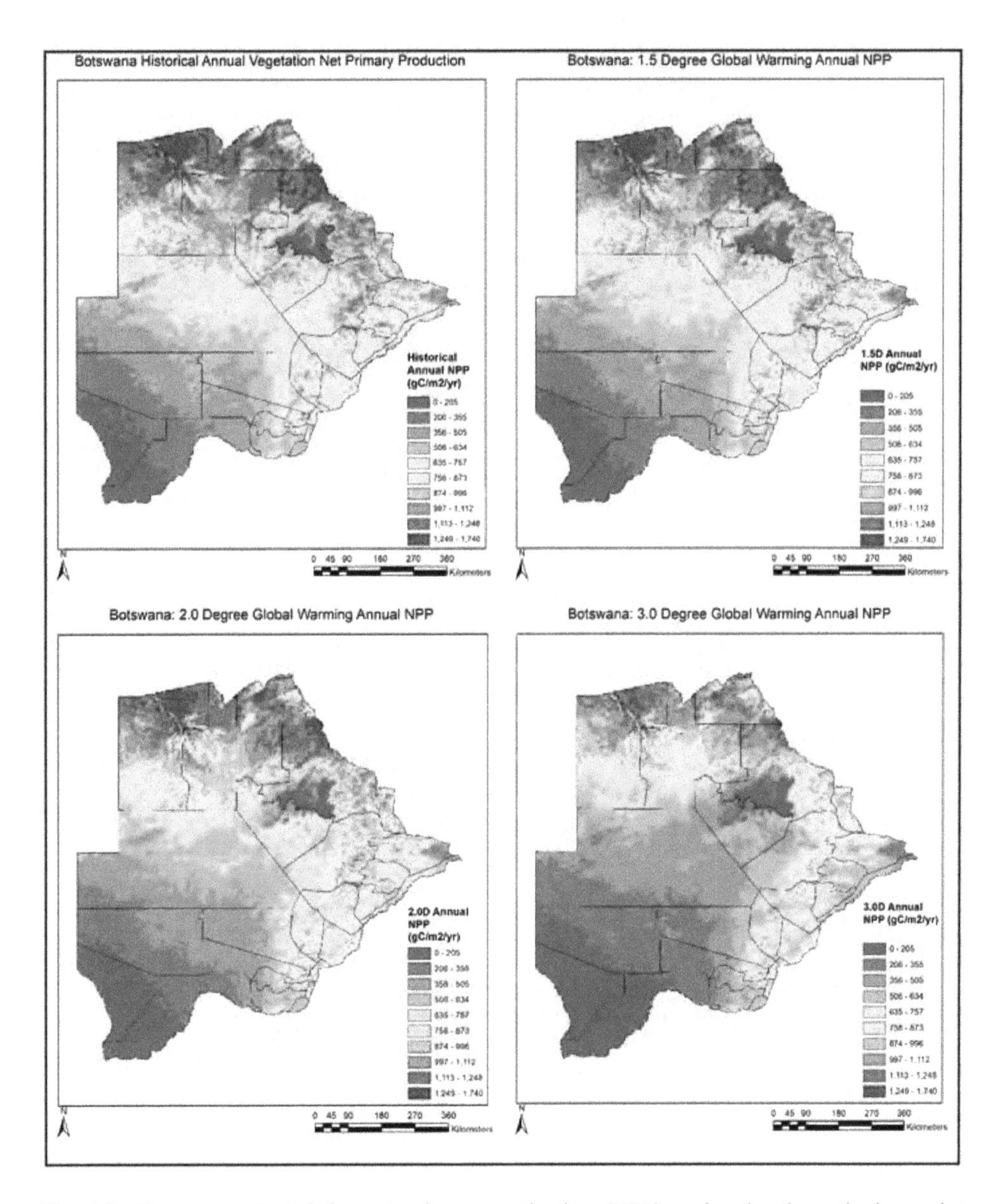

Fig. 10 Changes projected for net primary production (NPP) under the three tipping point scenarios of 1.5 °C, 2 °C, and 3 °C, from baseline (average of 2000–2016). (Source: authors)

will be required as the pace of climate-induced changes mirror those of society. The following needs have been identified:

(1) more specific, operational examples of adaptation principles that are consistent with unavoidable uncertainty about the future;
(2) a practical adaptation planning process to guide the selection and integration of recommendations into existing policies and programs; and,
(3) greater integration of social science into an endeavor that, although dominated by ecology, increasingly recommends extension beyond reserves and into human-occupied landscapes. (Heller and Zavaleta 2009: 14)

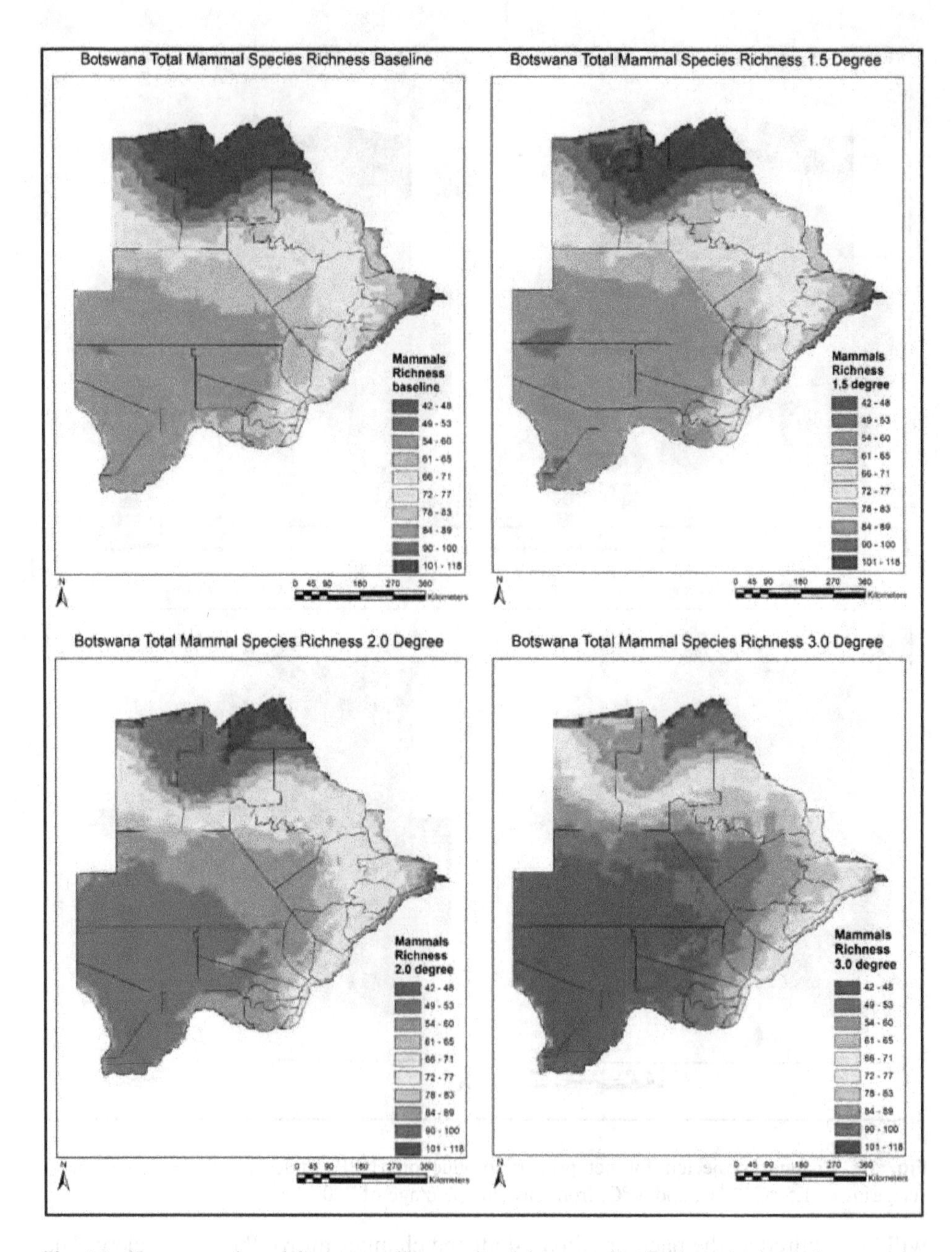

Fig. 11 Changes projected for mammalian species richness under the three tipping point scenarios of 1.5 °C, 2 °C, and 3 °C, from baseline. (Source: authors)

The Paris Agreement is deemed to be beneficial to the cause of species biodiversity conservation as it represents a template for a GDN by setting global targets, an evolving model for financial support, and recognizes the values obtained from bottom-up efforts. Nearly all nations have signed the agreement. Climate scientists have arrived at a single numerical target for maintaining Earth's atmosphere at safe

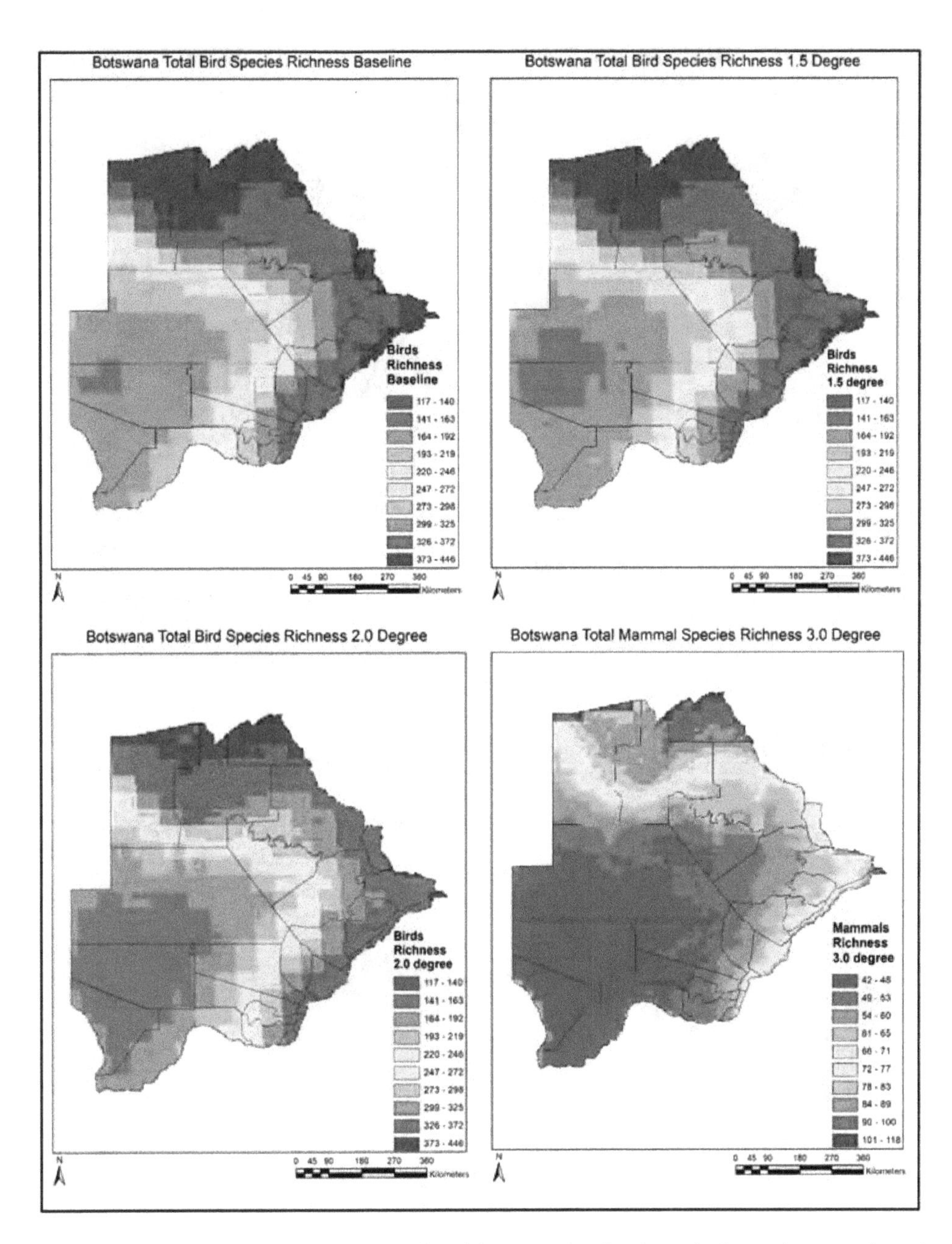

Fig. 12 Changes projected for bird species richness under the three tipping point scenarios of 1.5 °C, 2 °C, and 3 °C, from baseline (note: data source available is at a different resolution than mammalian and amphibian sources, hence the different map smoothness). (Source: authors)

limits (1.5 °C); however, biodiversity scientists work with multiple targets to conserve the rest of life on the planet given the diversity of life and ecosystems and range of resilience. The challenge clearly shown in the case of biodiversity in Botswana is the interconnectedness of the nation with the southern African region

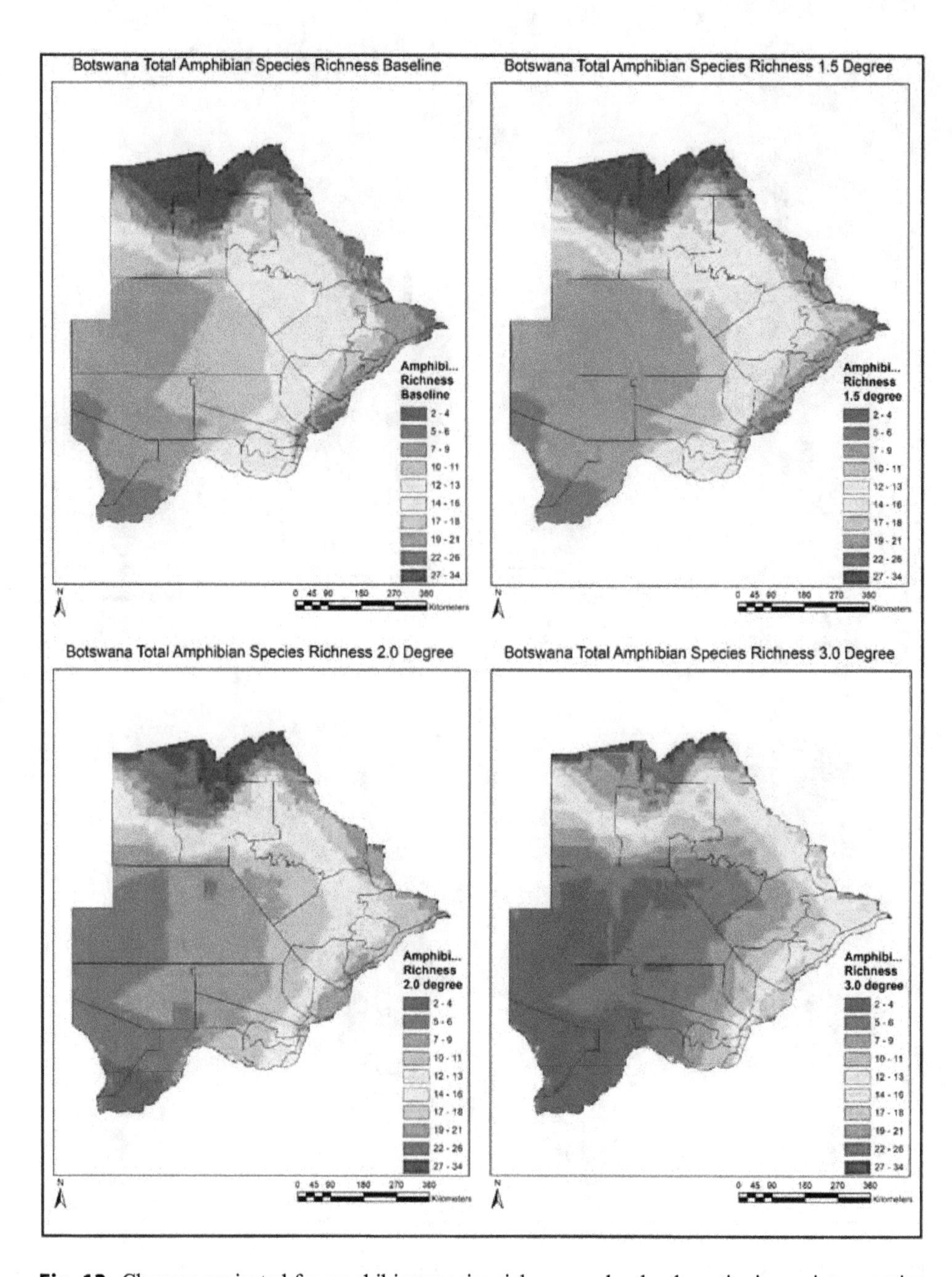

Fig. 13 Changes projected for amphibian species richness under the three tipping point scenarios of 1.5 °C, 2 °C, and 3 °C, from baseline. (Source: authors)

and the faster than global average rise in temperatures. The international climate science goal of limiting global temperatures to 1.5 °C by 2100 is not relevant for places such as Botswana that could reach that average rise in temperature as early as 2028. The GDN has expressed its own target like the Paris Agreement: protect at least half of Earth's biodiversity by 2050 and ensure that these areas are connected

Table 4 Biodiversity sector tipping point analysis summary table

Impact factors	Slow processes	Variabilities	Extremes	Potential tipping points
Earth systems (climate change): Threshold risks	Biodiversity decrease vegetation primary production, land degradation, access to water, heat stress, diseases, tourism and agriculture land pressure, conflict	International variability increase, death rate increase, diseases, poaching	Extreme drought and diseases (foot and mouth) events Low water flows in Okavango Delta Wildlife movements/ encroachments to cropping areas	Continuous drought years could cause the wildlife high death rate and migration
Earth systems (climate change): Potential adaptation options	Recover and reserve natural habitats for natural earth systems	Provide alternative areas/substantial buffer zones for ecosystem change and succession	System integrity enhancement to support biodiversity resilience at times of climatic stress	Species migration pattern shifts Species decline and extinction Increased human/ biodiversity conflict
Social systems: Threshold risks	Tourism sector dependency, livestock number increase, rural labor force, and food relationship	Tourism natural conservation and human conflicts	Wildlife decrease during an extreme drought event	Dramatic changes in tourist sector behavior Loss of economic tourism opportunity without changes in location and interaction with species
Social systems: Potential adaptation options	Tourists and local people education and behavior regulation	Natural and human relationship education and behavior regulation	Reduce human impacts, support system natural recovery where is possible	Biodiversity and tourism sector in collaboration with other ISEET for systematic analysis and uplift social system support

(continued)

Table 4 (continued)

Impact factors	Slow processes	Variabilities	Extremes	Potential tipping points
Institutional system: Threshold risks	Deficiency and out of date in biodiversity protection legislation and policy, land-use and planning legislation and policy	Government policy and institutional implementation capacity	Institutional dramatic changes	Policy failure
Institutional system: Potential adaptation options	Balanced policy and regulation and implementation capacity for biodiversity protection and tourism development	Policy and regulations support the resiliency of the biodiversity conservation that allows natural system recovery	Reinforce and stabilize institutional system, prepared for emergencies	Work together with closely linked ISEET systems, to build the robust institutions for biodiversity conservation and tourism
Economic systems: Threshold risks	Tourism sector over development Irrational land-use and agriculture development	International tourism market variations. Changed habitats and protected area systems too rigid for continuous protection of biodiversity as biomes shift	Market collapse, financial crisis	International financial crisis Tourist air travel depressed by GHG accounting and social pressure to reduce flying
Economic systems: Potential adaptation options	Financial regulation for overdevelopment and land development	Economic measures for sustainable tourism and biodiversity protection subsidies and compensation incentives	Tourism sector support and alternative work development	Overarching economic and financial arrangement for biodiversity and tourism, in coordination with ISEET systems

(continued)

Table 4 (continued)

Impact factors	Slow processes	Variabilities	Extremes	Potential tipping points
Technological systems: Threshold risks	Wildlife and biodiversity conservation technology improvement Human intervention at a large scale to enhance biodiversity management regimes	Technology insufficiency Maladaptation in the management of biodiversity and conservation	Technology failure Species collapse from linked with wider environmental collapse from extreme heat and drop in net primary production	Huge biodiversity loss by technological system failure Disease outbreak and biome changes lead to mass extinction events and loss of keystone species
Technological systems: Potential adaptation options	Sustainable knowledge and technology applications development and introduction for biodiversity and tourism	Development and introduction of the knowledge and technologies for resilient biodiversity and tourism sectors	Technology improvement and backup mechanisms	Technology innovation and knowledge gaining for ISEET system stabilization

(Dinerstein et al. 2019). Over time the role of intact, diverse systems has been repeatedly demonstrated, and they are integral to carbon storage and hence the link to the Paris Agreement. For Botswana, intactness and linkages regionally will become increasingly relevant and urgent given the accelerating pace of temperature and other changes facing the country, which are well ahead of the global trends (Table 4).

Summary and Conclusion

The implications of climate change are a growing concern for biodiversity managers in Botswana. Even simple extrapolation of current trends in temperature makes clear that a warming climate can have negative impacts on many facets of Botswana's society and ecosystems. The sum of these troubling effects, although highly uncertain, motivates efforts to assess and manage potential climate change impacts.

Climate change may trigger major social transformations. As seen around the globe, both slow-onset climate change and extreme weather events can, for example, induce food and water insecurity. Subsequent riots, mass movements of people, and pressure on governance systems to respond to limit social anxiety and loss of national cohesion are being increasingly researched.

However, it is not a simple, one-to-one relationship. The relationship between climate and biodiversity and social change is not easy to identify. The scale of climate change that can induce significant and consequential change and related calls to action, including social transformations, may be minor or major. Often it is built-in resiliency of Institutional-Socio-Economic-Ecological-Technological Systems (ISEETS) that determines if a drought, flood, fire, epidemic disease, or another extreme event will trigger a societal response.

Climate change experts such as Tim Lenton pioneered the use of the term tipping points to refer to "abrupt and irreversible changes in the climate system." The concept has been taken further than climate knowledge to further our understanding of the consequences of climate change. What are often overlooked in climate change research are the social determinants of changes, for example, inequities in determining the social impacts of climate change. Also ignored are perceptions and (mis) representations of change in whatever form that might be, as well as governance systems and institutions, solidarity networks, and cultural values, and underlying technological condition and opportunities in their evaluation of the future social impacts of climate change.

The experience of those interested in biodiversity in Botswana forms a recent historical example of how ecological tipping points are identified and sometimes managed historically ineffectively but with increased local and specific cultural and environmental awareness, more effectively. What are, therefore, more challenging to determine are the social tipping points that would potentially lead to more accurate assessments of the future impacts of climate change. This then expands the thinking about climate change to more fully grasp the biological effects of a changing climate and our technological capacity to forecast them and respond more effectively.

News has emerged on the recognition of the cross-sectoral importance of a more resilient Okavango ecological system which is encouraging. The signing in 2019 of the three-country pact to work, titled "Transboundary cooperation for protecting the Cubango-Okavango River Basin and improving the integrity of the Okavango Delta World Heritage," properly championing an integrated approach to the entire river basin management is an important step forward (Kari and Kaboza 2019). The actors and agents involved and their perspectives represent an ISEET-like approach. The follow-up activities of implementation of defining activities and their monitoring and evaluation will be critical. Time is running out as climate change and its impacts accelerate across Southern Africa.

References

Alexander J, Chase J, Newman N, Porter A, Roessner JD (2012) Emergence as a conceptual framework for understanding scientific and technological progress. In: 2012 Proceedings of PICMET'2: technology management for emerging technologies, pp 1286–1292

Anderson RL, Foster DR, Motzkin G (2003) Integrating lateral expansion into models of peatland development in temperate New England. J Ecol 91(1):68–76

Andrews P, O'Brien EM (2000) Climate, vegetation, and predictable gradients in mammal species richness in southern Africa. J Zool 251(2):205–231

Asayehegn K, Iglesias A, Triomphe B, Pédelahore P, Temple L (2017) The role of systems of innovation in adapting to climate change: the case of the Kenyan coffee and dairy sectors. J Innov Econ Manag 3:127–149

Baer HA, Singer M (2018) The anthropology of climate change: an integrated critical perspective. Routledge, London

Biggs R, Simons H, Bakkenes M, Scholes RJ, Eickhout B, van Vuuren D, Alkemade R (2008) Scenarios of biodiversity loss in southern Africa in the 21st century. Glob Environ Chang 18 (2):296–309

Ceccarelli A, Zoppi T, Vasenev A, Mori M, Ionita D, Montoya L, Bondavalli A (2019) Threat analysis in systems-of-systems: an emergence-oriented approach. ACM Trans Cyber Phys Syst 3(2):18

Christiansen L, Olhoff A, Trærup S (eds) (2011) Technologies for adaptation: perspectives and practical experiences. UNEP Risø Centre, Roskilde

Cole DH, Epstein G, McGinnis MD (2014) Toward a new institutional analysis of social-ecological systems (NIASES): combining Elinor Ostrom's IAD and SES frameworks. Indiana legal studies research paper, no. 299; Indiana University, Bloomington School of Public and Environmental Affairs research paper, no. 2490999. Indiana University, Bloomington School of Public & Environmental Affairs, Bloomington

Dai L, Vorselen D, Korolev KS, Gore J (2012) Generic indicators for loss of resilience before a tipping point leading to population collapse. Science 336(6085):1175–1177

Dinerstein E, Vynne C, Sala E, Joshi AR, Fernando S, Lovejoy TE, . . . Burgess ND (2019) A global deal for nature: guiding principles, milestones, and targets. Sci Adv 5(4):eaaw2869

Du K, O'Connor A (2019) Examining economic complexity as a holistic innovation system effect. Small Bus Econ. https://doi.org/10.1007/s11187-019-00215-z

Gerten D et al (2013) Asynchronous exposure to global warming: freshwater resources and terrestrial ecosystems. Environ Res Lett 8:034032

Goldstein JA (2018) Emergence and radical novelty: from theory to methods. In: Handbook of research methods in complexity science. Edward Elgar Publishing, Cheltenham

Gregory P, Stuart R (2013) The global economy and its economic systems. South-Western Cengage Learning, Boton, MA

Halley JD, Winkler DA (2008) Classification of emergence and its relation to self-organisation. Complexity 13(5):10–15

Heller NE, Zavaleta ES (2009) Biodiversity management in the face of climate change: a review of 22 years of recommendations. Biol Conserv 142(1):14–32

Hope M, McCloskey J, Hunt D, Crowley D, Bhloscaidh NM (2018) Innovation pathways to adaption for humanitarian and development goals: a case study of aftershock forecasting for disaster risk management. J Extreme Events 5(02n03):1850010

IPBES (2018) The IPBES regional assessment report on biodiversity and ecosystem services for Africa (eds: Archer E, Dziba L, Mulongoy KJ, Maoela MA, Walters M). Secretariat of the Intergovernmental Science-Policy Platform on Biodiversity and Ecosystem Services, Bonn. 492 pages. https://doi.org/10.5281/zenodo.3236178

IUCN (2008) Table 5: Threatened species in each country (Totals by Taxonomic Group)

IUCN (2019) Table 5: Threatened species in each country (Totals by Taxonomic Group)

Jones A (2017) Tourism in Botswana: tourist destinations in Botswana – wildlife. Amazon E-Book

Kari S, Kaboza Y (2019) Proceedings of the expert meeting "Transboundary co-operation for protecting the Cubango-Okavango River Basin and improving the integrity of the Okavango Delta World Heritage property", Maun, 3–4 June 2019. 36 pgs

Keys PW et al (2019) Anthropocene risk. Nat Sustain 2:667–673

Leadley P, Pereira HM, Alkemade R, Fernandez-Manjarrés JF, Proença V, Scharlemann JP, Walpole MJ (2010) Biodiversity scenarios: projections of 21st century change in biodiversity and associated ecosystem services. Secretariat of the Convention on Biological Diversity, Montreal. Technical Series no. 50, 132 pages

Lenton TM et al (2008) Tipping elements in the Earth's climate system. PNAS 105:1786–1793

Lenton TM, Livina VN, Dakos V, Van Nes EH, Scheffer M (2012) Early warning of climate tipping points from critical slowing down: comparing methods to improve robustness. Philos Trans R Soc A Math Phys Eng Sci 370(1962):1185–1204

Lichtenstein BB (2014) Generative emergence: a new discipline of organisational, entrepreneurial and social innovation. OUP US, New York

Liu Z, Otto-Bliesner BL, He F, Brady EC, Tomas R, Clark PU, ... Erickson D (2009) Transient simulation of last deglaciation with a new mechanism for Bølling-Allerød warming. Science 325(5938):310–314

Lucas K, Renn O, Jaeger C, Yang S (2018) Systemic risks: a homomorphic approach on the basis of complexity science. Int J Disaster Risk Sci 9(3):292–305

MacPherson AJ, Gillson L, Hoffman MT (2019) Between-and within-biome resistance and resilience at the fynbos-forest ecotone, South Africa. The Holocene 29(11):1801–1816

McCarthy NA (2003) Demand for rainfall-index based insurance: a case study from Morocco, vol 106. International Food Policy Research Institute, Washington, DC

McCord PF, Dell'Angelo J, Bladwin E, Evans T (2017) Polycentric transformation in Kenyan water governance: a dynamic analysis of institutional and social-ecological change: polycentric transformation in Kenyan water governance. Policy Stud J 45(4):633–658

Midgley GF, Thuiller W (2011) Potential responses of terrestrial biodiversity in Southern Africa to anthropogenic climate change. Reg Environ Chang 11(1):127–135

Newbold T (2018) Future effects of climate and land-use change on terrestrial vertebrate community diversity under different scenarios. Proc R Soc B Biol Sci 285(1881):20180792

Oliver TH, Marshall HH, Morecroft MD, Brereton T, Prudhomme C, Huntingford C (2015) Interacting effects of climate change and habitat fragmentation on drought-sensitive butterflies. Nat Clim Chang 5(10):941–945

Polhill JG, Filatova T, Schlüter M, Voinov A (2016) Modelling systemic change in coupled socio-environmental systems. Environ Model Softw 75:318–332

Porter AL, Garner J, Carley SF, Newman NC (2019) Emergence scoring to identify frontier R&D topics and key players. Technol Forecast Soc Chang 146:628–643

Roundy PT, Bradshaw M, Brockman BK (2018) The emergence of entrepreneurial ecosystems: a complex adaptive systems approach. J Bus Res 86:1–10

Ryan AJ (2007) Emergence is coupled to scope, not level. Complexity 13(2):67–77

Saarinen J, Hambira WL, Atlhopheng J, Manwa H (2012) Tourism industry reaction to climate change in Kgalagadi South District, Botswana. Dev South Afr 29(2):273–285

Schot J, Kanger L (2018) Deep transitions: emergence, acceleration, stabilisation and directionality. Res Policy 47(6):1045–1059

Steffen W, Rockström J, Richardson K, Lenton TM, Folke C, Liverman D, ... Donges JF (2018) Trajectories of the Earth System in the Anthropocene. Proc Natl Acad Sci 115(33):8252–8259

Sterner T, Barbier EB, Robinson A (2019) Policy design for the Anthropocene. Nat Sustain 2:14–21

Tàbara JD, Frantzeskaki N, Hölscher K, Pedde S, Kok K, Lamperti F, Berry P (2018) Positive tipping points in a rapidly warming world. Curr Opin Environ Sustain 31:120–129

United Nations Framework Convention on Climate Change (UNFCCC) (2005) Report on the seminar on the development and transfer of technologies for adaptation to climate change: note by the Secretariat. FCCC/SBSTA/2005/8

van der Vleuten E (2019) Radical change and deep transitions: lessons from Europe's infrastructure transition 1815–2015. Environ Innov Soc Trans 32:22–32

Waisman H, Bataille C, Winkler H, Jotzo F, Shukla P, Colombier M, ... La Rovere E (2019) A pathway design framework for national low greenhouse gas emission development strategies. Nat Clim Chang 9(4):261

Walby S (2007) Complexity theory, systems theory, and multiple intersecting social inequalities. Philos Soc Sci 37(4):449–470

Wilson F, Moser G, Fallon D, Farrell CA, Müller C, Wilson D (2019) Rewetting degraded peatlands for climate and biodiversity benefits: results from two raised bogs. Ecol Eng 127:547–560

Winterbach HE, Winterbach CW, Somers MJ (2014) Landscape suitability in Botswana for the conservation of its six large African carnivores. PLoS One 9(6):e100202. https://doi.org/10.1371/journal.pone.0100202

Winterbach CW, Whitesell C, Somers MJ (2015) Wildlife abundance and diversity as indicators of tourism potential in Northern Botswana. PLoS One 10(8):e0135595. https://doi.org/10.1371/journal.pone.0135595

8

Gendered Vulnerability to Climate Change Impacts in Selected Counties in Kenya

Daniel M. Nzengya and John Kibe Maguta

Contents

Abstract

Extreme climate change events such as frequent and prolonged droughts or floods associated with climate change can be very disruptive to peoples' livelihoods particularly in rural settings, where people rely on the immediate environment for livelihood. Shocks in the people's livelihoods can trigger diverse responses that include migration as a coping or adaption strategy. Migration takes many forms

D. M. Nzengya (✉)
Department of Social Sciences, St Paul's University, Limuru, Kenya
e-mail: dmuasya@spu.ac.ke

J. K. Maguta
Faculty of Social Science, St Paul's University, Limuru, Kenya

depending on the context and resources availability. Very few studies in Kenya have used qualitative analysis to bring up women's voices in relation to gender, climate change, and migration, especially along hydrological gradient. This chapter presents results of qualitative research conducted from 58 participants in 2018 in three counties in Kenya, namely, Kiambu County, Machakos, and Makueni. The study sought to examine gender perceptions related to climate-induced migration, that is: whether climate change is perceived to be affecting women's livelihood differently from that of men; examine in what ways experiences of climate induced migration differed for men and women; explore perceptions on the county government efforts to cope with climate-induced migration; and examine perceptions of the role of nongovernmental agencies in helping citizens cope with climate change. From the results obtained on ways in which climate change affected women livelihoods more than men had four themes: (1) women exerted more strain in domestic chores, child/family care, and in the farm labor; (2) women also experienced more time demands. The sources of water and firewood were getting more scarce leading to women travel long distances in search to fetch water and firewood; (3) reduced farm yields, hence inadequate food supply; and (4) the effects of time and strain demands on women was a contributory factor to women poor health and domestic conflicts. Several measures that the county government could take to assist women to cope with climate change-induced migration had five themes which include the following: (1) developing climate change mitigations, and reducing deforestation; (2) increasing water harvesting and storage; (3) develop smart agriculture through the use of drought-resistant crops and drought mitigation education; (4) encourage diversification of livelihoods; and finally (5) providing humanitarian assistance to the most vulnerable populations such as orphans and the very poor. Thirdly, the measures mentioned that NGO's could take to assist rural communities to cope with climate change-induced migration did not vary significantly from those mentioned for county government, except probably for a new theme of *increasing* advocacy for climate adaption policies.

Keywords

Climate change-induced migration · Hydrological gradient · Gendered vulnerabilities · Rural livelihoods · Kenya

Introduction and Background

Climate change poses one of the greatest challenges to rural livelihoods in the Sub-Saharan Africa (SSA). Sub-Saharan Africa is among the regions of the world where majority of people still reside in the rural areas. Millions of households in the rural populations rely on rain-fed small-scale subsistence agriculture (Kalungu et al. 2013). Subsistence agriculture has continued to face many sustainability challenges in the last 10 years, namely, worsening land degradation, declining farm outputs due to declining parcel of land associated with continuing subdivision due to rising population, inaccessibility to farm drought resistance seed crops, extreme climate change events, particularly more frequent and prolonged droughts (Kalungu et al. 2013). Because rural population has risen during the last 15 years, increasing demands for farming and settlement has pushed communities to marginal and ecologically fragile lands, for instance, steep hill slopes, river banks, arid and semiarid areas (Ngugi et al. 2015). Poor land use practices have further exacerbated the continent' problem of soil erosion and desertification (Ngugi et al. 2015), further compounding population's vulnerability to extreme climate events such as droughts. It is estimated that climate change is partly responsible for the rural-urban migration in the SSA region, as people flee to towns and cities to pursue alternative livelihoods (Barrios et al. 2006; Hassan and Tularam 2018).

According to the Intergovernmental Panel on Climate Change, there is need to put in place mitigation and adaptation measures (Intergovernmental Panel on Climate Change 2018). It is because of climate change threats to human life that the United Nations took quick steps that resulted into holding conventions to mitigate climate change effects. According to the United Nations Framework Convention on Climate Change (UNFCCC) report developed from workshops held in 2006–2007 in Africa, Asia, and Latin America, all nations and people have a stake in reversing climatic changes (United Nations Framework Convention on Climate Change 2006). The UNFCCC paints a worrying report with claims that the developing countries are feeling the effects of climate change more than the developed ones. While developed countries have better and advanced coping mechanisms, the developing ones rely on crude means that leave them suffering the wrath of nature emanating from drought, excessive heat, and floods, and so on (Hurd and Smith 2004). There is need to put in place mitigation and adaptation measures as well as doing all that can be done to ensure the ecosystem remains balanced and conserved.

The objectives of this chapter are to: (1) examine if climate change was affecting women's livelihood differently from that of men; (2) in what ways climate-induced migration differed for men and women; and (3) to examine the measures county government and nongovernmental agencies were taking assist communities cope with climate-induced migration.

Overview of Global Perspectives of Climate-Induced Migration

Extreme climate change events interact with many factors to trigger human migration. For instance, the rising trends in climate related disasters, such as severe and prolonged droughts, floods, in different regions of the world have forced people to migrate for safety or for survival. Van der Land and Hummel (2013) has argued that there has been a tendency of low rainfall forcing man to degrade the land. When land is degraded, the environment is affected even more. Further, planning for people has become harder. It is hard to allocate resources to populations that keep moving from one region to another when threatened by the same weather the populations have contributed to make unfriendly. When populations migrate, they further strain the resources of their new areas of settlement. Kartiki (2011) has expressed concern that the ensuing migrations are likely to increase conflicts and political stability. This, the author argues, will result from the struggle over the scarce resources and probably reshaping of geographical boundaries. International conflicts are likely to increase when countries share boundaries.

Chindarkar (2012) has lamented that climate change though painful to all leaves the female gender suffering more in comparison to men. The author argues that women are disadvantaged in terms of property ownership and getting education making them poorer. Failure to give women good education and allow them access to natural resources leaves them with little options to deal with hazards of climate change among other outcomes that include poor health and forced migrations. There are reduced economic activities that are based on natural resources resulting in food insecurity among other things. Climate change induced displacements of people as well as migrations and relocation have continued to take place making the UN to take climate change as an agenda that calls for urgent action (United Nations 2016). Several agencies including the United Nations have called upon organizations and individual countries to have measures in place to address displacement and migration of people (United Nations 2016). However, despite the calls and efforts of the UN, migrations and displacements of people have been on the increase (Hassan and Tularam 2018).

Population movements occasioned by climate change are feared to adversely affect the poor and the vulnerable that have low coping mechanisms. Chindarkar (2012) has opined that since adaptation to and coping with climate change are gendered, women will continue to suffer as they have little income and access to natural resources in comparison to their male counterparts. Reduced access to food and water lays a bigger burden on women. As evidenced in the case of Chitwan Valley in Nepal, women are in most cases the primary collectors of the various provisions required by family members. Citing the case of Sonora in Mexico, the author states that the reduction of water affected the food processing industry forcing men to migrate to towns. The outcome was increase in the workload of women who were left with the burden of tending for families without the men who had migrated. In a similar trend, internal migrations have taken place quite to a large extent in the republics of Kiribati and Tuvalu as a result of induced climate change (Hassan and Tularam 2018). Majority of the people especially women in the Kiribati and Tuvalu rely almost entirely on agricultural-related activities.

The gender divide puts a heavier burden on women who are tasked with looking for water and much more often searching for food. The socially constructed gender differences heap the obligations and the burden of caregiving on women. Since the economic muscle needed to acquire food and water is in many cases lacking, as men monopolize this, women are left with no option but to trek long distances looking for the commodities. Kolmannskog (2009) in a study conducted in Somalia and Burundi has stated that women who are left behind by their husbands face another fear of being chased away by families or relatives of their husbands. When husbands migrate in search of better pastures for their animals, women and children are left behind with relatives.

This has been the trend in Mexico where apart from destroying plants and aquatic life, health has been affected through water pollution and sedimentation. Studies conducted in Bihar and Uttarakhand in India have confirmed that the people who have limited access to resources are more vulnerable and suffer more from climate change effects. Adding their voices to the debate on climate change-induced migrations, Waldinger and Fankhauser (2015) have opined that it increases incentives to migrate. Inability to cope with induced climate change has called for emergency relief and security for those affected (Serdeczny et al. 2017). The authors further argue that internal displacements have resulted in greater risks especially sexual and gender-based violence for women. The latter authors have claimed that when people have to move as a result of weather, the dressing code has compounded the problem like what happened to the Indian women during the Indian Ocean tsunami in 2004. The mode of dressing for Indian women is not suited for walking fast and covering long distances. Further, climate-induced migrations and displacements often lead to breaking of social networks and psychological impacts that are lasting. The authors here have argued that women feel the impact of climate change-induced migrations more deeply. In many cases, they are left with no choice but to think of themselves as well as their entire families. The situation becomes even worse for women as they are rarely allowed direct access to relief food and other emergencies due to cultural constrains. The scenario portrays a potential risk of women falling victims of sexual exploitation or even worse to be trafficked. In countries like Bangladesh and Philippines, due to their low education, women are subjected to low-working paying jobs with long working hours. They are employed as domestic house-helps who have to contend with mistreatments in order to earn bread. Climate-induced migration is not only dangerous but has also turned into a poverty trap (Rahman 2013).

Climate change has forced people to adapt into new cultures in order to survive. Chindarkar (2012) has cited the case of Bangladesh women who have migrated to India. For fear of detection and eventual deportation, many have adorned Hindu religious markers especially on their foreheads.

In Mali and Senegal both of which are in Africa, climatic changes have adversely affected subsistence farming and livestock rearing. Since this Sahel region of West Africa has majority of the people relying on the rains that have been declining, migration has become a common phenomenon (van der Land and Hummel 2013). With a good number being illiterate or semiliterate, farming has been the only way to as they cannot be absorbed in gainful employment that require skills. Waldinger and Fankhauser (2015) have claimed that people involved in the agricultural sector in developing countries will continue suffering climate shocks like droughts and flooding (also, see IPCC 2018). These are attributed to weak financial muscle and use of low technology.

Climate-Induced Migration in Kenya: Causes and Implications

Like the rest of the world, effects of climate change are being felt in Kenya. This is proved by the erratic and unpredictable weather patterns. Cuni-Sanchez et al. (2019) content that the amount of rainfall, fog, and temperature has witnessed serious variations in the last few decades. Rain seasons have changed making it hard to predict the weather and prepare land in good time for farmers. With the high reliance on rain-fed agriculture, variability of rains has threatened food security and complicated lives mainly in rural areas. Droughts and flooding are now a common thing with crops being destroyed. Desertification has increased making life even tougher especially for pastoralists whose livestock have faced grass shortages. Livelihoods have been destroyed with vulnerability increasing by the day. Migrations mainly to urban centers for wage employment especially by men have been taken as the option by some of the affected. Like in other countries, women have been left behind to cater for the families (IPCC 2018).

Climate variability in Kenya like the rest of the world has been blamed largely on human activities. As Sheikh (2017) has argued in a study of Dadaab area of North Eastern Kenya, human-related activities have led to stiff competition over resources which in turn has led to increased conflicts. The increased conflicts have compounded [the problem created the?] movement of people from one place to another. Further, the internally displaced persons in the country are exposed to other vices that include sexual and gender-based violence, dependency, instability, and living in fear as the future remains uncertain. With the scarce resources in the country, migrations emanating from climate change are a major threat to the well-being of all as one is either directly or indirectly affected. Sabbarwal (2017) has lamented on increasing temperatures in Turkana region with conflicts increasing as people compete over reducing pasture and water resources. The author claims that with the area drying up and with temperatures having gone up by approximately 2 °C between 1967 and 2012, raids and migrations have been common. With the rainy seasons becoming shorter and drier, women and girls who are charged with the duties of fetching water have no option but to trek long distances to get the diminishing resource. Water has to be extracted by digging dry wells and riverbeds. With the reduced pastures, animals have died or become famished.

Cuni-Sanchez et al. (2019) carried out a study in northern Kenya and noted a lot of climatic changes in Mt. Kulal, Mt. Nyiro, and Mt. Marsabit. Natural rivers had turned into seasonal rivers. Reliance on firewood and other non-timber products had been affected greatly by changing climate. The unreliability of rains that at times would come when not expected and in low quantities forced people to device different coping mechanisms which include migrations.

Climate-induced migrations have complicated life for those affected and created problems to the government (Corburn et al. 2020). Corburn and others (2020) recent work have pointed to Kenya's urban slums to a haven for COVID-19 infections. Some demographic projections estimate that over half of citizens in Kenya's capital, Nairobi reside in the slums (UN-Habitat n.d.). This has made it hard to share resources like water and housing. Both the national government and the county government have to struggle in provision of public goods and services to the slum population that is increasing by the day. Security has become an issue while environmental health has become an eyesore with dirt in the congested settlements becoming a nuisance.

Materials and Methods

This study employed use a qualitative cross-sectional survey design. Data were collected in 2018 from three counties using questionnaire with open-ended questions. The open-ended questions included: (1) if climate change was affecting women's livelihood differently from that of men; (2) ways in which climate change-induced migration of women differed from those of men; (3) the initiatives the county government should undertake to help people cope with climate change; and finally, (4) the participants' perceptions of the measures nongovernmental agencies should undertake to assist communities cope with climate change.

Study Setting

The research was conducted at three counties in Kenya which are Kiambu, Machakos, and Makueni counties. The Kiambu county is generally lies at a higher attitude, is generally cool with high amounts of rainfall, followed by Machakos, with Makueni being classified as arid and semiarid county. The Kiambu County is generally wet, and the average annual rainfall for the Kiambu is 1000 mm annually. Machakos County, on the other hand, receives average rainfall, with some parts of the county experiencing similar weather patterns as Kiambu County, and other parts being relatively dry. Makueni County is classified as an arid and semiarid region. The county receives low amount of annual rainfall of 600 mm. Makueni county experiences frequent and prolonged droughts triggering crop failures, and is among the list of counties in that receives relief food to mitigative starvation, malnutrition.

Data Analysis and Presentation of Results

This study used content analysis (Hsieh and Shannon 2005) and inductive coding (Strauss and Corbin 1998). Content analysis is a "technique for making inferences by objectively and systematically identifying specified characteristics of messages" (Holsti 1969, p. 14). Content analysis allows a blend of both quantitative and qualitative data analysis attributes to be combined. That is, the researchers can identify count data and compute frequencies/percentages for further analysis. In addition, these quantitative measures can be supported with themes/categories. This method was adopted due to the short nature of responses that was generated from the survey open-ended questions. There were two coding cycles. The first cycle enabled the researchers to describe the data. In the second coding cycle, the codes identified in the first cycle were compared, organized, and categorized (Tracy 2013). To increase credibility of the coding procedure, two people were involved, one of the researchers and a colleague who is familiar with the subject matter. The two researchers did both coding cycles together and where there was a disagreement, they discussed until they reached an agreement. Where possible thick descriptions were identified to support the themes/categories identified. In this study, the unit of analysis was the respondent and each of the four question were analyzed separately to identify themes.

Results and Discussions

Climate Change Affects Women's Livelihood Differently from That of Men: Empirical Evidence from the Study

Twenty two percent of the respondents did not think climate affects women's livelihoods differently than men. However, 88% of the respondents felt climate affected women's livelihoods differently from men. Table 1 summarizes the different themes emerged from the analysis of qualitative responses to the question "in what ways

Table 1 Ways in which climate change affected women livelihood differently from that of men

Themes	Subthemes	Explanation	No of mentions	%
More strain	Domestic chores	Women faced more strain as most of the household chores were assigned to them. For instance, collecting firewood and fetching water	4	6.45
	Child/family care	Women faced more strain from taking care of the young ones	5	8.06
	Farm labor	Women had more responsibilities in preparing the farm, cutting pasture for animals, and harvesting	6	9.68
Low income	Low yields and inadequate food supply	The farms yielded low yields, leading to inadequate food supply for the family and poor nutrition	8	12.90
	Women as bread earners	Women had to run small agro-related businesses/farming to supplement or basically provide for the family	10	16.13
Time demands	More time demands searching water and firewood	Women had to walk long distances searching for firewood and water	10	16.13
	More time demands searching food	Women had to walk long distances searching for food	3	4.83
Effects of time and strain demands	Poor health	Women were prone to sicknesses due to effects of harsh cold weather in the fields, and strain and experienccd cold-related sicknesses such as colds, asthma, arthritis, among others	14	22.58
	Domestic conflicts	The strain and time pressures contributed to domestic conflicts between the women and their spouses	2	3.22
				100

Source: Authors' survey results, 2020

does climate change affect women's livelihoods differently from that of men." Four themes, namely, more strain, low income, time demands, and effects of time and strain demands. The theme "more strain" can be described as "exert extra effort in carrying out domestic and farm related chores." This theme was comprised of four themes described in Table 1, namely, domestic shores, child/family care, and farm labor. Women in the rural areas in many parts of the SSA, particularly in Kenya, bear disproportionate burden of household chores (Muasya and Martin 2016; Mokomane 2014). As men migrate to towns and cities to look for better income opportunities, this adds to the burden women bear as a consequence of climate change particularly prevalent and prolonged droughts. Majority of rural areas lack access to reliable and/ or affordable water sources, and most households rely on surface water resources for household and domestic needs (Kelly et al. 2018). Most of these sources are seasonal rivers and water springs which run dry during dry spell, meaning women have to walk longer distances during drought seasons to look for the scarce commodity (Cherutich et al. 2015). As one of the participants A1 remarked:

> women are more burdened as they have to source for their families' necessities

while participant B1 had the same concerns:

> Women travel long distances searching for water for domestic use apart from caring for the family

Also, unlike urban areas where working class women hire house helps to assist with child care and other family chores (Muasya 2014), women in rural areas grapple with insufficient resources for survival, meaning they have to bear the burden of child and family care demands. Previous studies have shown that where husbands are present, they provide some support in these tasks reducing stress related to child/ family care and work-balance conflicts (Muasya 2016). However, as men migrate to towns and cities, rural women have to bear the extra burden of attending to family chores while carrying children on their backs. As one participant B2 commented:

> Women are more affected as they have to fetch water and take care of the young ones

It is estimated that rural women provide over 80% of farm labor in the rural areas at the SSA region (Ogunlela and Mukhtar 2009). Majority of the rural households lack resources for mechanization, meaning they rely on rudimentary tools and approaches to accomplish most of the farm activities, from digging, land preparation, planting, weeding, harvesting, and postharvest tasks (FAO 2011). As men and young people migrate to towns and cities to run-away from difficulties precipitated by climate change in rural areas, which means women have to bear the burden of accomplishing all these tasks on their own as participant A5 alluded:

> Women are more into the farms harvesting crops and also its preparation

The theme "low income" implies that women earned low income due to declining yields. This theme comprised two subthemes, namely, low yields and inadequate food supply, and low income from agro-related businesses. Rural women rely on selling a variety of farm produce for financial sustenance, and also to provide support to diverse family needs, from purchasing food to educating their children, farm produce consists the main supply of food and nutrition to families. In a study of stressors and work family conflict among urban female teachers, Muasya (2020) found that low income was a stressor to teachers with low income which further made them seek extra sources of income exacerbating their work life balance challenges. In addition, severe and prolonged droughts associated with climate change have been prevalent in the study sites, particularly Makueni County in the last 5 years, resulting into massive crop failures and livestock deaths (Amwata et al. 2015). This led to massive starvation, malnutrition, especially where humanitarian interventions are rarely available (Amwata et al. 2015). This is worsened as most men migrating to towns and cities end up in low-income informal employment and lack extra income to sustain their rural families left behind in rural areas. As one participant C5 remarked:

> Yes. Most of women depend on small scale farming which due to climate change do not produce enough

This was echoed by participant C8,

> Low yields make women struggle very much to find food

The theme "time demands" implies "women spent more time in household chores, farming than men who migrate to towns for industrial jobs." This theme comprised of two subthemes, namely, more time demands searching water and firewood, and more time demands searching food. Although for households living in urban areas, both men and women participate in paying water bills, energy bills, and purchasing food, on the contrast, in the rural areas in most parts of SSA region, women are the ones responsible for collecting water, firewood for cooking, looking for food and the cooking as well (Tian 2017). The burden of the multiple task of looking for food and the time expended walking long distances to look for water particularly during droughts, collecting firewood causes real strain and stress upon rural women (Tian Tian 2017). The quotes by two survey participants below highlight the predicament rural women face out of the cxperiences during extreme climate change events, particularly droughts. As participants A5, B3, and C4 remarked, respectively:

> women spent a lot of time searching for clean water
> Yes. Women waste a lot of time walking far distances looking for water unlike men
> Women walk long distances searching for water hence wasting a lot of time

The last theme identified was on the effects of time and strain demands. The results of the survey identified two subthemes that had to do with how participants perceived climate change events to affect women, namely, poor health and domestic conflicts. Many participants attributed the vulnerability of women to women's prolonged hours in the farms, or walking to collect water and/or firewood to increase the prevalence of illnesses such colds, asthma, arthritis, among others. Given that women have to carry along their infants during most of these tasks, participants were concerned that their infants and children were by extension exposed to the negative health effects experienced by rural women due to climate change.

Besides poor health, women experienced domestic conflicts. Too much strain and time pressures led to strained marital relationships between the spouses. The women could not be in a position to be equally present at home to take care of children, prepare food in good time, or even offer conjugal rights to their spouses. Studies have shown women even in the formal sector with no adequate support for child care tend to quit their work (Muasya 2017). On the other hand, delegation of house chores and childcare chores can be a source of conflict as well, when these women fail to undertake these chores as socially expected. Moreover, as more household and farm tasks are carried out by women than their spouses, these women can perceive it that their spouses have neglected their duties and it can result to more domestic strife. Indeed, lack of work life balance and delegation of chores can be a source of domestic conflict (Muasya and Martin 2016).

Ways in Which Experiences of Climate Change-Induced Migration Differed for Men and Women

Majority of participants (90%) felt that that climate change-induced migration differed for men and women. Specifically, more men than women left home and migrated from rural areas to urban areas to look for alternative sources of income, search for pasture, or work in farms. Consequently, women were left with more responsibilities. The quotes below highlight participants perceptions on the different ways in which experiences of climate-induced migration differed for men and women as participant A10 narrated:

> men migrate to look for jobs while women engage in simple chores to cater to-day-today needs

and participant A12 remarked:

> most men migrate in search of alternative sources of livelihood while women remain behind to cope with the change

And participant B6 said:

> diverse climatic change makes most men travel to towns leaving families behind

Measures by County Governments to Assist Communities Cope with Climate Change-Induced Migration

Table 2 summarizes the different themes emerged from the analysis of qualitative responses to the question "measures by county governments to assist communities cope with climate change induced migration." Eight themes, namely, climate change

Table 2 Measures by county governments to assist communities cope with climate change-induced migration

Themes	Subthemes	Explanation	No of mentions	%
Climate change mitigation/ adaptation		Encourage reforestation, planting of trees, agroforestry, and reduce cutting of trees	7	6.31
Women empowerment		Women empowered to make financial decisions that can empower them	6	5.40
Strengthening water adaptive capacity	Increased water harvesting and storage	One way to curb drought is increased water harvesting storage through provision of water harvesting resources such as funds, materials, e.g., water tanks, involved in sinking boreholes, dam reservoirs	47	42.34
	Sensitization on water harvesting	Women farmers to be sensitized on the need of water harvesting and ways to harvest water	3	2.70
Soil and water conservation		The need to preserve soil and water through measures such as gabions, water friendly trees, curbing farming along rivers, etc.	8	7.21
Climate smart agriculture	Drought-resistant crops	Farmers to grow drought-resistant and fast-maturing crops with minimal irrigation	15	13.51
	Drought education	Sensitization and education on drought-tolerant crops and information on weather patterns, e.g., from meteorological department	8	7.21
Diversifying livelihoods		Women to be encouraged to explore other non-farm income-generating activities	6	5.40
Humanitarian assistance		The county and nongovernmental organizations to identify and support the most vulnerable groups to climate change such as orphans and the very poor	5	4.50
Individual and institutional capacity building		Enhancing the capacity of women and the local institutions on ways to harvest and conserve water	6	5.40
				100

Source: Authors' survey results, 2020

mitigation and adaptation through tree planting; women empowerment; strengthening households' water adaptative capacity; soil and water conservation (resilience capital); promoting climate smart agriculture; sharing meteorological information/data with rural households; promoting diversification of livelihoods; and targeted humanitarian assistance and strengthening individual and institutional capacity.

The theme on tree planting did not have many categories, and there was little variability of responses from sampled participants. Suggested items included "reforestation"; "sensitizing people to plant trees"; "educating residents on importance of planting trees"; "planting drives on water catchment"; "encouraging people to plant more trees"; "encourage people to avoid cutting down of trees"; and "encouraging tree nurseries." There are several ways in which these measures mentioned by participants can be linked to coping with climate change-induced migration. Planting trees becomes a source of income from the sales of animal fodder, firewood, fruits, thereby providing women with the need supplies to meet households and domestic needs. Some fruits planted in the study sites, particularly Machakos and Makueni Counties, include grafted mangoes that incidentally do well during seasons of crop failures, thereby acting as a buffer to household income and food supplies (Muema et al. 2018).

Women empowerment was another theme that emerged from the analysis of participants' responses regarding the open-ended question on what measures by county governments to assist communities to cope with climate change-induced migration. In most rural settings in the SSA region, patriarchal structures largely influenced decisions related to access to land for cultivation, farm inputs, harvest and postharvest, access to capital to support farm labor, access to capital for physical assets such as farm equipment, water harvesting, and improved cooking stoves (FAO 2011). Even where microfinance existed to support women's efforts, many women had to still seek consent from their spouses, even when the needs were pressing. Such patriarchal structures further limit or frustrates women's intention to build asset portfolio (social networks, human capital, natural capital, physical capital, and financial capital) necessary for climate resilience, and adaptation. The last decade or so has seen many countries including Kenya address women's marginalization particularly in relation to land inheritance, land ownership, and this has spillover benefits particularly reducing gendered vulnerabilities associated with climate change events such as droughts (FAO 2011).

The theme on strengthening households adaptative capacity has several subthemes, namely, increasing supply of materials for water harvesting/funds; supplying water tanks; drilling boreholes; digging dams and reservoirs; and sensitization of communities on water harvesting. Women and children, especially in the rural areas, suffer most problems associated with inaccessibility to reliable and affordable safe sources of water for household and domestic use (Graham et al. 2016). Consequently, improving water accessibility greatly improves the capacity of rural households to cope with climate induced migration, particularly when they are faced with limited human capital because men/husbands have migrated to towns and cities. Also, in pastoral communities, climate change-induced conflicts over diminishing water sources and pasture are quite common in the SSA (Witsenburg and Adano

2009). Consequently, measures such as digging boreholes, water harvesting structures such as dams help communities to better cope with induced migration as they have access to more options in terms of water sources to support their livestock (Witsenburg and Adano 2009).

Soil and water conservation was another theme that was identified, and three subthemes, namely, gabions, water friendly trees on catchment areas, and avoiding farming near rivers. Soil erosion and bad land use practices such cultivating along rivers backs undermine the resilience capital rural dwellers desperately require coping with climate change-induced migration.

Climate smart agriculture is further identified to be another theme that emerged in the analysis of qualitative responses related to county government measures to assist communities to cope with climate-induced migration. Three subthemes identified here included: drought-resistant or fast-maturing crops/seeds; irrigation; and sensitization and education of drought-tolerant crops. Other themes identified included: sharing meteorological information/data with rural households; diversifying livelihoods; timely and strategic humanitarian assistance; and strengthening rural institutions.

Measures by Nongovernmental Agencies to Assist Communities Cope with Climate Change-Induced Migration

Six themes emerged from the qualitative analysis of participants' responses in relation to measure nongovernmental agencies have taken to assist communities cope with climate change-induced migration. Four of the themes seemed to overlap with measures undertaken by county governments. However, two of these themes differed significantly from the measures participants suggested in relation to the role of the county government, namely, climate change advocacy and diverse range of humanitarian interventions. Statements implying climate change advocacy were most frequently mentioned measures by respondents.

The range of humanitarian interventions that NGOs were involved with to assist communities cope with climate change and consequences of climate-induced migration included: supporting children from very poor families; providing food to poor families; providing medical care, providing mobile clinics for health; child sponsorship; providing cheaper energy options for cooking; and providing seeds for planting (Table 3).

Lessons Learned, Study Limitations, and Recommendations for Future Research

Rural households in the study sites observed that climate change affects women's livelihoods differently from men. Specifically, climate change exerts more strain on women, lowers their agro-dependent income due to crop failures and reduced yields, increases time demands women have to spend on household chores such as looking

Table 3 Measures by nongovernmental agencies to assist communities cope with climate change-induced migration

Themes	Subthemes	Explanation	No of mentions	%
Adaptation through water harvesting	Providing water harvesting resources	This water harvesting can be achieved through provision of water harvesting resources and assistance in sinking boreholes and dams	7	8.33
	Water harvesting education and sensitization	Provide water harvesting education and sensitization to communities	2	2.38
Climate change advocacy		Engage the communities and government on policy change to change to new ways of farming technologies	21	25%
Support livelihood diversification			10	11.90
Support climate mitigation measures			13	15.48
Climate smart agriculture			16	19.05
Humanitarian intervention			15	17.86
				100

Source: Authors' survey results, 2020

for water as local water sources dry up, looking for firewood, and the effects of time and strain demands, and have negative impacts on women's, infants, and children's health. Rural communities felt that climate change-induced migration differed for men and women; specifically, more men than women left home and migrated from rural areas to urban areas to look for alternative sources of income, search for pasture, or work in farms. Consequently, women were left with more responsibilities. Regional/county governments' measures to assist communities cope with climate change-induced migration include: supporting climate change mitigation and adaptation through increased tree planting; women empowerment; strengthening households' water adaptative capacity; soil and water conservation (resilience capital); promoting climate smart agriculture; sharing meteorological information/data with rural households; promoting diversification of livelihoods; targeted humanitarian assistance; and strengthening individual and institutional capacity. While nongovernmental agencies also participate in most of these measures, they seem to bring in strength in relation to climate change advocacy and a very diverse range of humanitarian interventions.

The study sample was limited the three counties in Kenya, and findings may not be generalizable in other parts of the SSA. The rural livelihoods communities studied are more-or-less sedentary, and future studies may need to focus on pastoral/nomadic communities. Also, the study investigated gendered vulnerabilities to climate change impacts and climate-induced migration largely among rural communities' settings. Future research will need to expand this work to urban settings and coastal communities.

The research design followed was cross-sectional survey, future studies are needed to include large sample, and probably use longitudinal designs to observe gendered vulnerabilities and climate-induced migration over a relatively longer study period.

In this study, trained research assistant participants filled responses to the open-ended questions according to participants answers. This limited the ability to capture participants emotions that can add perspective to the qualitative data collected. Future cross-sectional surveys would need to record participants narration, transcribe verbatim, the responses then analyze data.

Conclusion

Indeed, we can conclude that climate change does have negative effects on the livelihoods of rural women in a more disproportionate way compared to men. Thus, intervention whether through the county government or nongovernmental organizations may be required to factor in gender and socioeconomic factors in their policies and intervention programs. Failure to mitigate the negative effects of climate change in rural settings might worsen the livelihoods of women and children, undermining progress the sustainable development of reducing gender inequality, and also health for all citizens.

References

Amwata DA, Nyariki DM, Musimba NR (2015) Factors influencing pastoral and agro-pastoral household vulnerability to food insecurity in the drylands of Kenya: a case study of Kajiado and Makueni counties. J Int Dev 28(5):771–787

Barrios S, Bertinelli L, Strobl E (2006) Climatic change and rural-urban migration: the case of sub-Saharan Africa. J Urban Econ 60:357–371

Cherutich J, Maitho T, Omware Q (2015) Water access and sustainable rural livelihoods: a case of Elementaita division in Nakuru county, Kenya. Int J Sci Technol Soc 3(1):9–23

Chindarkar N (2012) Gender and climate change-induced migration: proposing a framework for analysis. Environ Res Lett 7(2):025601

Corburn J, Vlahov D, Mberu B et al (2020) Slum health: arresting COVID-19 and improving well-being in urban informal settlements. Yale School of Medicine. https://medicine.yale.edu/news-article/23734/. Accessed 5 July 2020

Cuni-Sanchez A, Omeny P, Pfeifer M, Olaka L, Mamo MB, Marchant R, Burgess ND (2019) Climate change and pastoralists: perceptions and adaptation in montane Kenya. Clim Dev 11(6): 513–524

FAO (2011) The State of Food and Agriculture 2010–11 | FAO | Food and Agriculture Organization of the United Nations. Available at: http://www.fao.org/publications/sofa/2010-11/en/. Accessed 25 June 2020

Graham JP, Hirai M, Kim SS (2016) An analysis of water collection labor among women and children in 24 sub-Saharan African countries. PLoS One 11:e0155981

Hassan OM, Tularam GA (2018) The effects of climate change on rural-urban migration in sub-Saharan Africa (SSA) – the cases of democratic Republic of Congo, Kenya and Niger. In: Applications in water systems management and modeling, vol 10. IntechOpen, London, pp 64–68
Holsti OR (1969) Content analysis for the social sciences and humanities. Addison-Wesley, Reading
Hsieh HF, Shannon SE (2005) Three approaches to qualitative content analysis. Qual Health Res 15 (9):1277–1288
Easterling WE III, Hurd BH, Smith JB (2004) Coping with global climate change: the role of adaptation in the United States. Pew Center on Global Climate Change, Arlington
Tollefson, J. IPCC says limiting global warming to 1.5 C will require drastic action. Nature 2018, 562, 172–173.
Kalungu JW, Leal Filho W, Harris D (2013) Smallholder farmers' perception of the impacts of climate change and variability on rain-fed agricultural practices in semi-arid and sub-humid regions of Kenya. J Environ Earth Sci 3:129–140
Kartiki K (2011) Climate change and migration: a case study from rural Bangladesh. Gend Dev 19 (1):23–38
Kelly E, Shields KF, Cronk R, Lee K, Behnke N, Klug T, Bartram J (2018) Seasonality, water use and community management of water systems in rural settings: qualitative evidence from Ghana, Kenya, and Zambia. Sci Total Environ 628–629:715–721
Kolmannskog V 2009 Climate change, disaster, displacement and migration: initial evidence from Africa New Issues in Refugee Research No. 180 (United Nations High Commissioner for Refugees)
Mokomane Z (ed) (2014) Work–family interface in sub-Saharan Africa: challenges and responses. Springer, Cham
Muasya G (2014) The role of house helps in work–family balance of women employed in the formal sector in Kenya. In: Work–family interface in sub-Saharan Africa. Springer, Cham, pp 149–159
Muasya G (2016) Work-family balance and immigrant sub-Saharan women in the United States. In: Communication and the work-life balancing act: intersections across identities, genders, and cultures. Lexington Books, London, pp 163–184
Muasya G (2017) Work-family conflict, support and intention to quit among Kenyan female teachers in urban public schools. S Afr J Labour Relat 41(1):33–45
Muasya, G. (2020). Stressors and work-family conflict among female teachers in urban public schools in Kenya. South African Journal of Education, 40(2).
Muasya G, Martin JN (2016) Conflict in Kenyan households: an exploratory study of professional women and domestic workers. Howard J Commun 27(4):385–402
Muema E, Mburu J, Coulibaly J, Mutune J (2018) Determinants of access and utilisation of seasonal climate information services among smallholder farmers in Makueni County, Kenya. Heliyon 4 (11):e00889
Ngugi L, Rao K, Oyoo A, Kwena K (2015) Opportunities for coping with climate change and variability through adoption of soil and water conservation technologies in semi-arid Eastern Kenya. In: Adapting African agriculture to climate change. Springer, Cham
Ogunlela YI, Mukhtar AA (2009) Gender issues in agriculture and rural development in Nigeria: the role of women. Humanit Soc Sci J 4(1):19–30
Rahman MS (2013) Climate change, disaster and gender vulnerability: a study on two divisions of Bangladesh. Am J Hum Ecol 2(2):72–82
Ravera F, Martín-López B, Pascual U, Drucker A (2016) The diversity of gendered adaptation strategies to climate change of Indian farmers: a feminist intersectional approach. Ambio 45 (Suppl 3):335–351
Sabbarwal S (2017) Indigenous peoples' concerns for environment: examining the role of non-governmental organizations. Fourth World J 15(2):27–39
Serdeczny O, Adams S, Baarsch F et al (2017) Climate change impacts in sub-Saharan Africa: from physical changes to their social repercussions. Reg Environ Chang 17:1585–1600
Sheikh AA (2017) Effects of climate variability on human migration dynamics: a case study of Ifo refugee camp, Daadab complex – Kenya, p 110

Strauss A, Corbin JM (1998) Basics of qualitative research techniques and procedures for developing grounded theory. Sage, Thousand Oaks

Tian X (2017) Ethnobotanical knowledge acquisition during daily chores: the firewood collection of pastoral Maasai girls in Southern Kenya. J Ethnobiol Ethnomed 13:2

Tracy SJ (2013) Qualitative research methods. Wiley-Blackwell, Chichester

UN-Habitat (n.d.) Kenya Habitat Country Program Document: 2018–2021 enhancing effective service delivery and sustainable urban development at national and county levels. UNHabitat, Nairobi

United Nations (2016) General adopts declaration for refugees and migrants as United Nations Organization sign key agreement. Retrieved from https://www.un.org/press/en/2016/ga11820.doc.htm

United Nations Framework Convention on Climate Change handbook (2006) UNFCCC handbook. UNFCCC Climate Change Secretariat, Bonn

van der Land V, Hummel D (2013) Vulnerability and the role of education in environmentally induced migration in Mali and Senegal. Ecol Soc 18(4):14

Waldinger, M., & Fankhauser, S. (2015). Climate change and migration in developing countries: evidence and implications for PRISE countries.

Witsenburg KM, Adano WR (2009) Of rain and raids: violent livestock raiding in Northern Kenya. Civ Wars 11(4):514–538

9

Underutilized Indigenous Vegetables' (UIVs) Business in Southwestern Nigeria: Climate Adaptation Strategies

V. A. Tanimonure

Contents

Abstract

The impact of climate change, especially on agricultural sector, calls for a global and more localized strategies such as cultivation of underutilized indigenous vegetables (UIVs) which adapt better to local climate change. This chapter, therefore, examines the perception of UIVs farmers to climate change, their experiences of UIVs' responses to climate change, adaptation strategies

V. A. Tanimonure (✉)
Agricultural Economics Department, Obafemi Awolowo University, Ile-Ife, Nigeria
e-mail: tanimonurevic@oauife.edu.ng

employed, and the determinants of the decision to adopt them in Southwest Nigeria. The study uses quantitative and qualitative primary household data from 191 UIVs farmers, 8 Focus Group Discussions (FGDs), and secondary climate data from the Nigerian Meteorological Agency. Descriptive and econometric analyses are employed in the data analyses. The results show that farmers' perceptions of climate change are high temperature and a high variability in rainfall pattern that has affected the yield, increased insects, pests, and diseases infestations, and reduced soil fertility. The results further show that the responses of UIVs to these resultant effects differ as such, and adaptation strategies farmers adopt are UIVs-specific. The adaptation strategies mostly employed by the UIVs farmers are cultivating UIVs along the river bank and the least is agroforestry and perennial plantation. The determinants of the decision to adopt adaptation strategies include UIVs revenue, age, years of experience, access to climate information, climate change awareness, agro ecological zone, and access to credit. Thus, promotion of UIVs business is advocated and provision of information on climate change essential and will encourage farmers to adopt appropriate climate change adaptation strategies to boost UIVs business.

Keywords

Climate change · Adaptation strategies · Underutilized · Indigenous vegetables · Nigeria

Introduction

The impact of climate change forecast on agriculture has posed a threat to the sustainability of global food security and nutrition. For instance, over half a billion (525 million) people in the tropics were projected by Consortium of International Agriculture Research (CGIAR) program on Climate Change Agriculture and Food Security (CCAFS) to possibly be at the perils of hunger by 2050 due to climate change (Actionaid International 2011). Intergovernmental Panel on Climate Change (IPCC) (2014) also predicted climate change increasing current problems and also generating new ones for natural and human systems. Likewise, Boko et al. (2007) predicted that reductions in yield of up to 50% in some countries in Africa by 2020 and net crop revenue by as much as 90% by 2100 could be attributed to climate change, smallholder farmers being mostly affected, especially the rural farmers who depend largely on rain-fed agriculture (Ching and Stabinsky 2011). Until adaptive or palliative measures are inaugurated to mollify the effects of climate change, food security in developing countries in the tropics, particularly in Nigeria, will be under threat (Enete 2014), considering the high rate of population growth.

Owing to the undesirable effects of climate change, it is suggested that a blend of global and more localized strategies, along with other adaptation strategies, can help farmers weather the effects of climate change. These include conservation agriculture, organic agriculture, carbon sequestration and the capacity to withstand weather stresses, change in planting time, the breeding of a number of climate-resilient crop varieties such as underutilized indigenous vegetables (UIVs), among others (Howden et al. 2007; Omatseye 2009; Sambo 2014). Underutilized indigenous vegetables are vegetables that originate from a locality; such vegetables may be localized to that particular area or found in other places. They are reported to be grown more widely or intensively in the past but are falling into disuse for a number of agronomic, genetic, economic, and cultural reasons. They are promising species but their potentials in terms of economic, nutrition, medicinal, and resilience have not been fully harnessed. Farmers, marketers and consumers are not making the best use of these crops as much as they use others because they are less competitive compared with other crops in the same agricultural environment (Guarino 1997; Eyzaguirre et al. 1999; IPGRI 2002; Padulosi and Hoeschle-Zeledon 2004). More so, these vegetables have not been a subject of organized research until recently (Tanton and Haq 2008). For instance, in the recent times, the production, marketing, and consumption of some of these vegetables are promoted in Southwest Nigeria in projects tagged NiCanVeg and MicroVeg. Some studies also reported that these UIVs are more adaptive, resilient, and tolerant to adverse climatic conditions than exotic species (Raghuvanshi and Singh 2001; Nnamani et al. 2009; Mabhaudhi et al. 2016).

In comparison to other crops, vegetables generally are more susceptible to environmental extremities such as high temperatures and soil moisture stress. Carbon dioxide, a major greenhouse gas, influences their growth and development as well as incidence of insect pests and diseases that render vegetable production unprofitable (Devi et al. 2017; Abewoy 2018). And all these will increase in the face of climate change (Ayyogari et al. 2014). However, UIVs can be produced comparatively at lower management cost, on marginal soil, and can tolerate the dynamics of climate change better than the exotic vegetables (Raghuvanshi and Singh 2001; Nnamani et al. 2009; Padulosi et al. 2011). This offers a significant opportunity for poor people in the rural areas to withstand the effect of climate change and increase their revenue, and, thereby, be food and nutrition secure (Maroyi 2011; Ebert 2014). This is because many UIVs from the tropics are already well adapted to the climatic conditions, and, as such, respond better to climatic conditions.

Unfortunately, there is limited quantitative and qualitative information supporting all these claims (Padulosi et al. 2011; Intergovernmental Panel on Climate Change (IPCC) 2014; Chivenge et al. 2015). The knowledge of UIVs adaptation to climate change remains concealed in the indigenous knowledge systems and this may explain why certain communities have continued to preserve and utilize certain UIVs. The limited quantitative empirical information indicates that UIVs remain under-researched as well. Although, some studies identified the potentials of these neglected and underutilized crops in sub-Saharan Africa in the recent times. A research by Adebooye and Opabode (2004), for example, was particular about the conservation of these indigenous crops in order to prevent them from going into extinction, especially as the reliance on a handful of major crops has innate agronomic, nutritional, and economic risks, which is not sustainable in the long run (Ebert 2014). Ayanwale et al. (2011) and Aju et al. (2013) also studied the market potentials of some of these underutilized indigenous vegetables and envisaged good commercial prospect in them, especially for women folks who are resource constrained. Some of the representatives of these leafy vegetables, tuber crops, cereals, and grain legumes that fit into the class of underutilized crops were identified by Maroyi (2011), Shrestha (2013), and Chivenge et al. (2015). It was equally found out that they are potential future crops for smallholder farmers, as sources of nutrition and income, especially in this era of climate change. Study by Sambo (2014) showed that underutilized crops could offer scientists a rich source of genetic materials for modification, which could hold potential key to developing resilient and drought-tolerant crops. Recent research found out that these underutilized indigenous crops have the ability to grow under water-scarce conditions and that the key to future food and nutrition security may lie in their untapped potentials (Mabhaudhi et al. 2016). More recent research shows qualitatively that climate change has both positive and negative effects on indigenous vegetables and the predicted negative effect cannot be overemphasized (Chepkoech et al. 2018). Indigenous vegetables farmers, therefore, adopt a number of adaptation strategies to mitigate the effect of climate change on their production activities. Although Fadairo et al. (2019) examined the perceived likelihood impact and adaptation of vegetables farmers to climate change, their emphasis was not on indigenous vegetables which is the focus of this study. Their study advocated for locality-specific climate change adaptation strategies.

The Intergovernmental Panel on *Climate Change* (IPPC) defines *adaptation* as "any adjustment in natural or human systems in response to actual or expected *climatic* stimuli or their effects which moderates harm or exploits beneficial opportunities." Climate adaptation is termed as a correct adjustment to climate variability and change, especially for smallholder farmers to enhance resilience or reduce vulnerability to its effects (Fadairo et al. 2019).

This study, therefore, aims to examine thc perception of UIVs' farmers and the responses of UIVs to climate variability and change in the study area. It assesses the adaptation strategies adopted by the farmers to cope with the adverse effects of climate change and lastly, analyzes the determinants of the decision to adopt adaptation strategies by the UIVs farmers. This will add to the existing literature on the resilience of underutilized indigenous vegetables to climate variability and change in sub-Saharan Africa.

Methodology

The Study Area

The study area is South west region of Nigeria as presented in Fig. 1. The area lies between longitude 2° 31[1] and 6° 00[1] east and latitude 6° 21[1] and 8° 37[1] norths. The study area is bounded in the east by Edo and Delta States, in the north by Kwara and Kogi States, in the west by the Republic of Benin, and in the south by the Gulf of Guinea. The region constitutes about one sixth (~163,000 km^2) of the total land area of Nigeria and comprises six states (Oyo, Ogun, Osun, Ondo, Ekiti, and Lagos) and is distinctly divided into three major agro-ecological zones (Rain Forest zone, Swamp Forest zone, and Derived Savannah zone) with diverse climatic conditions. The forest agro-ecological zone has annual rainfall in the range of 1,600–2,400 mm, with cropping seasons between April and November, while dry spells occur from December to March. The soil types in this zone depend largely on parent rock; where the underlying rocks are granite or clay, the soil is a rich clayey loam. On the other hand, the derived savannah agro-ecological zone has mean annual rainfall ranging from 800 to 1,500 mm, with cropping seasons between June and November. The soil types range from the sandy to clayey in texture, with soil reaction ranging from acidic to slightly basic. Soil fertility statuses and crop species diversity also vary widely in different locations in the region. This study is carried out in two of the three agro-ecological zones: Rain Forest and Derived Savannah zones, where UIVs were promoted in a project tagged NiCanVeg.

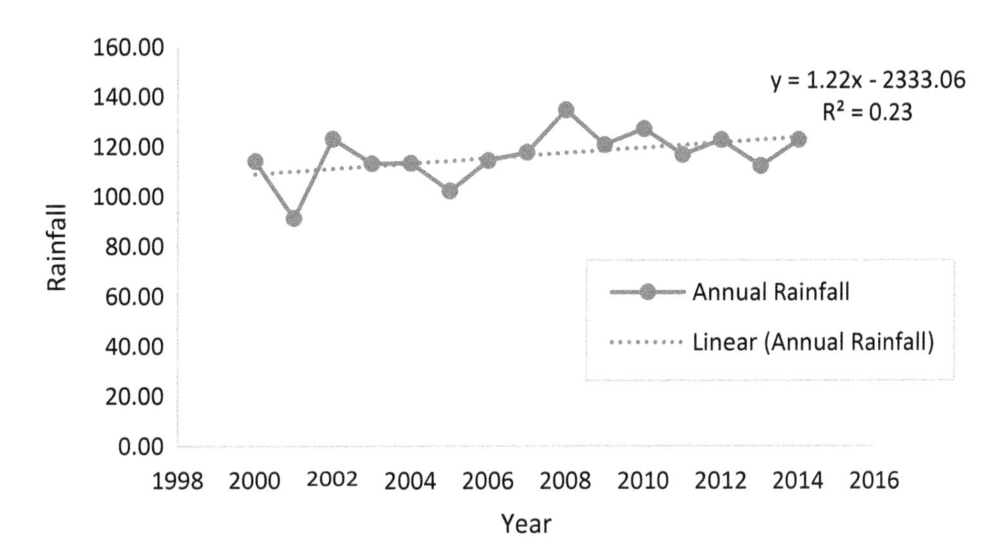

Fig. 1 Annual rainfall trend in Southwest Nigeria

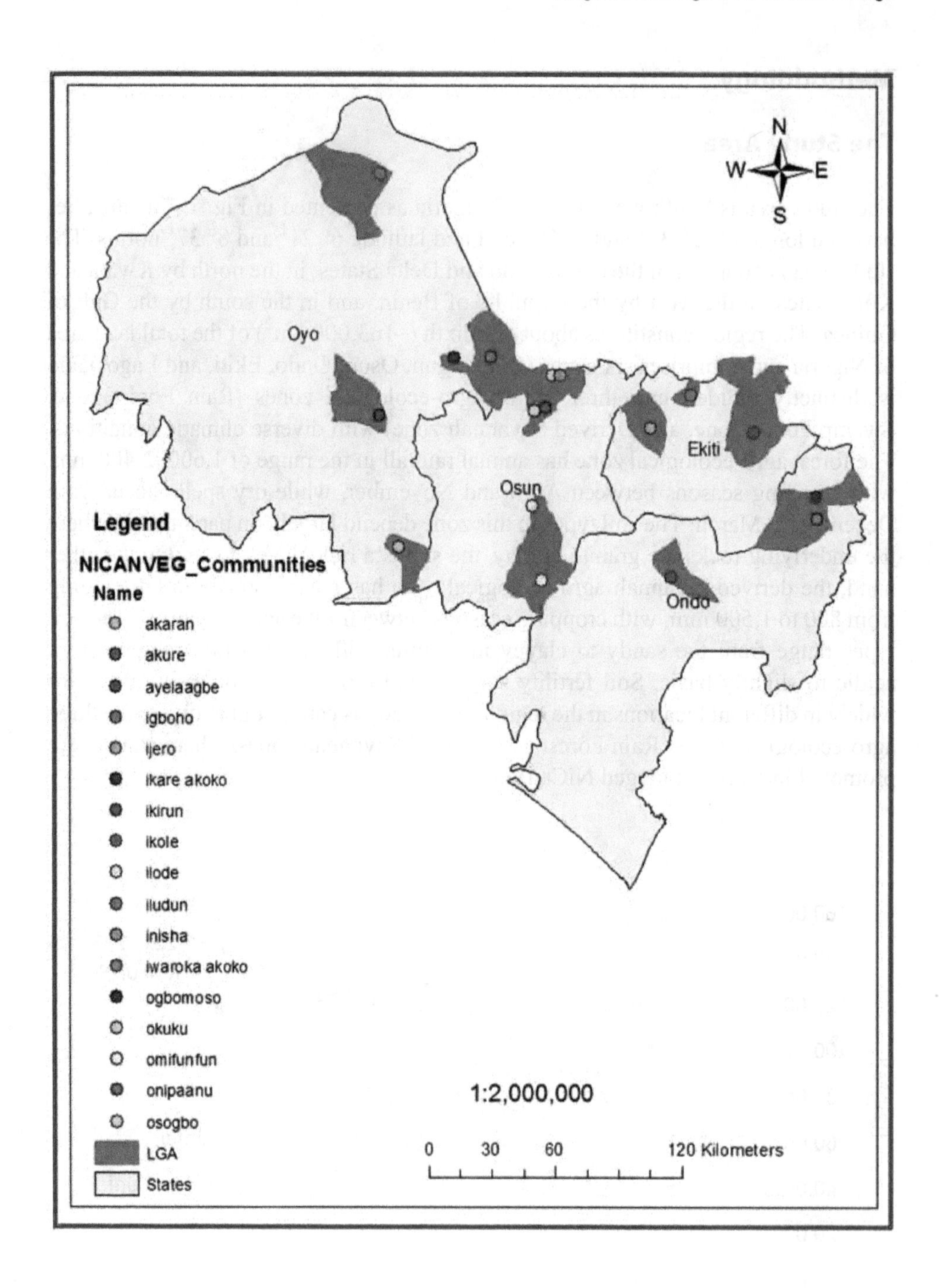

Sample Size and Sampling Procedure

All the 17 NiCanVeg communities in 16 Local Government Areas (LGAs) of the four (Ekiti, Ondo, Osun, and Oyo) states were selected for this research. In order to ensure representativeness and due to limited budget, a simplified formula (Eq. 1) developed by Kothari (2004) was used to calculate the sample size of the respondents at the community level. A 95% confidence level, 5% estimated percentage, and $P = 0.5$ were assumed in the equations.

$$n = \frac{Z^2 X\, p\, X\, q\, X\, N}{e^2 X\, (N-1) + Z^2 X p\, X\, q} \tag{1}$$

where

n is the sample size
N is the population size
e is the estimated proportion
p is sample proportion, $q = 1 - p$
z is the value of the standard variate at a given confidence level and to be worked out from Table showing area under Normal Curve.

Based on this formula, the respondents' sample size is approximately 191 (which was about 50% of the direct beneficiaries of NiCanVeg project in the study area).

In each NiCanVeg site, the NiCanVeg farmers were stratified into male and female to ensure random selection of both sexes, and 50% of the total farmers were randomly selected from NiCanVeg farmers' lists. This proportionate sampling procedure was necessary because of the difference in the number of farmers in each community or site.

Data

The mixed-methods research design involving both the quantitative and qualitative research approaches were employed to elicit information from the respondents. The quantitative study involved face-to-face data collection with the use of structured questionnaire. The questionnaire administration was done by the trained enumerators to ensure the quality of the data. Before the data collection, there was an "advance notification" sent to the respondents to let them know that the survey would be conducted in their communities. The enumerators were led to the communities by the NiCanVeg field officer who was already familiar with the farmers and the communities. Two Focus Group Discussions (FGDs) per state were conducted among the UIVs farmers to gather the qualitative data used for the study. The data collected include UIVs' household socioeconomic characteristics, their perception of climate change, and the various adaptation strategies adopted over the years to mitigate the effects of climate change, information on various vegetables they cultivate, the reasons why they cultivate the vegetables, and the responses of the vegetables to the perceived negative effects of climate variability in the study area. The FGDs with the farmers generated information on the type of vegetables cultivated across the four states, their order of economic importance, and the responses of these vegetables to resultant or indirect effects of climate change variables (rainfall and temperature) such as drought, flood, insect infestations, pests, and diseases. Farmers were asked to describe climate change variables expressed as changes in amount of rainfall and temperature on the extent of drought, changes in insect infestations, soil fertility, and UIVs yield. Also, UIVs producers' perceptions of climate change are compared with historical trends from Nigerian Meteorological data (average monthly temperature and rainfall) between 2000 and 2014.

Data Analysis

Both descriptive statistics and econometric analyses are used in the data analysis. Descriptive statistics such as frequency counts, means, and percentages are employed to describe the socio-economic characteristics of the respondents and the effects of climate change adaptation strategies on UIVs production. Content analysis was employed for the qualitative data collected through the FGDs. For the econometric analysis, logistic regression inferential statistic was used to analyze the factors affecting UIVs farmers' decisions to adopt adaptation strategy or not (Mahouna et al. 2018). In this study, the dependent variable is dichotomous, that is, farmers' decision to either adopt or not adopt climate change adaptation strategy. This method is appropriate because it considers the relationship between a binary dependent variable and a set of independent variables.

Results and Discussion

Table 1 summarizes the distribution of the UIVs that farmers cultivate in order of economic importance across the four States that the study covered. The summary reveals while good number of the UIVs are found across the entire region, only few UIVs are state specific. For instance, it is only in Osun State that red amaranth is produced in commercial quantity. Also, it is only in Oyo State (northern part) that *Solanum zuccagnianum* (locally called osun) is cultivated in commercial quantity. Aside Ondo State where ugu is the most economically important UIV, amaranth species remain the most economically viable UIV in the region. It is also noteworthy that respondents in Oyo State ranked two different UIVs as first economic important vegetables. While the UIVs producers in the northern part of the State ranked *Solanum zuccagnianum* as number one economically important UIV, those in the southern part ranked amaranth species as the number one in term of economic importance.

Farmers' Perception of Climate Change

Perception is the way something is regarded, understood, or interpreted. It is one of the first important steps in the process of designing some form of change in farmers' livelihood system to adapt to the changing climate. In order to get essential information and insight into farmers' perception of climate change, the two most important

Table 1 Distribution of UIVs production in the study area

Local name	English name	Scientific name	Order of economic importance
Osun State			
Tete Abalaye	White amaranth	*Amaranth viridis*	1
Red Tete/Tete Ijesa	Red amaranth	*Amaranth cruentus*	2
Ewedu	Jute mallow	*Corchorus olitorius*	3
Ugu	Fluted pumpkin	*Telfairia occidentalis*	4
Igbagba/Gboma	African eggplant	*Solanum macrocarpon*	5
Worowo	Bologi	*Solanecio biafrae*	6
Soko	Quail grass	*Celosia argentea*	7
Waterleaf	Waterleaf	*Talinum fruticosum*	8
Ebolo	Fire weed	*Crassocephalum crepidoides*	9
Elegede	Field pumpkin	*Cucurbita moschata*	10
Ogunmo	Garden huckleberry	*Solanum scabrum*	11
Oyo State			
Osun	–	*Solanum zuccagnianum*	1
Tete Abalaye	White amaranth	*Amaranth viridis*	1
Ogunmo	Garden huckleberry	*Solanum scabrum*	2
Ewedu	Jute mallow	*Corchorus olitorius*	3
Igbagba/Gboma	African eggplant	*Solanum macrocarpon*	4
Soko	Quail grass	*Celosia argentea*	5
Ugu	Fluted pumpkin	*Telfairia occidentalis*	6
Odu	Black nightshade	*Solanum nigrum*	7
Ebolo	Fire weed	*Crassocephalum crepidoides*	8
Ondo State			
Ugu	Fluted pumpkin	*Telfairia occidentalis*	1
Igbagba/Gboma	African eggplant	*Solanum macrocarpon*	2
Tete Abalaye	White amaranth	*Amaranth viridis*	3
Soko	Quail grass	*Celosia argentea*	4
Elegede	Field pumpkin	*Cucurbita moschata*	5
Ogunmo	Garden huckleberry	*Solanum scabrum*	6
Worowo	Bologi	*Solanecio biafrae*	7
Odu	Glossy nightshade	*Solanum nigrum*	8
Ekiti State			
Abalaye	White amaranth	*Amaranth viridis*	1
Igbagba/Gboma	African eggplant	*Solanum macrocarpon*	2
Ugu	Fluted pumpkin	*Telfairia occidentalis*	3
Ewedu	Quail grass	*Celosia argentea*	4

(continued)

Table 1 (continued)

Local name	English name	Scientific name	Order of economic importance
Waterleaf	Fire weed	*Crassocephalum crepidoides*	5
Odu	Field pumpkin	*Cucurbita moschata*	6
Soko	Quail grass	*Celosia argentea*	7
Worowo	Bologi	*Solanecio biafrae*	8
Ogunmo	Garden huckleberry	*Solanum scabrum*	9

Source: Field survey, 2016

Table 2 Farmers' perception on rainfall and temperature change

Farmers' perception	Rainfall	Temperature
No change	10 (5.24)	10 (5.24)
Yes, increasing	8(4.19)	161(84.29)
Yes, decreasing	134(70.16)	14(7.33)
Erratic	38(19.90)	0
Indifference	1(0.52)	6(3.14)

Source: Field survey, 2016

elements of climate rainfall and temperature are considered in this study. Out of the 191 sample respondents, only 10 (5.24%) are of the opinion that there is no change in the climate as presented in Table 2. This result shows that farmers are well aware of climate change. This result is similar to Fadairo et al. (2019), who also found that awareness of climate variability and change was high among vegetables farmers.

Farmers' Perception of Change in Rainfall Versus Meteorological Data

The perception of UIVs farmers in the study area on the overall trend of average rainfall as presented in Table 2 shows that all the respondents who are aware of climate change perceived changes in the rainfall, although their perception of these changes differs. About 74% of the respondents perceived decrease in average rainfall; about 4% perceived increase in rainfall, while about 21% could not say categorically whether the rainfall increased or decreased, but noticed an erratic rainfall over the years. Only 1% are indifferent to the rainfall pattern. The trend analysis of rainfall from Meteorological data between 2000 and 2014 in the area under study is presented in Fig. 1. The trend shows that there is no particular trend in average annual rainfall, as the rainfall pattern has been erratic. In year 2000, the average annual rainfall is high and falls in 2001. In 2002, the rainfall increases, falls in 2003, remains a little steady in 2004, and falls in 2005. The highest average rainfall within the period under study is in 2008 and since then, the average annual rainfall keeps rising and falling, although the quantitative trend shows that there is increase in average annual rainfall amount in the study area. The result of regression analysis between rainfall and time shows that an increase in one-year period results in a corresponding increase in the amount of average annual rainfall by 1.22 mm (Fig. 1). However, the perception of most farmers of rainfall shows a view contrary to the information contained in the meteorological recorded data. The majority of UIVs producers perceived reduction in rainfall. This lack of congruence could be as a result of the farmers assessing rainfall in relation to the needs of UIVs at a particular time; small change in quantity, onset, and cessation of rain over days make a big difference in the hearts of farmers, whereas the Meteorological data is more likely to measure total and large effects (Lemmi 2013).

Farmers' Perception of Change in Temperature Versus Meteorological Data

In the case of average temperature, all the UIVs farmers who are aware of climate change perceived changes in average temperature over the years (Table 2). Majority (88.95%) of UIVs producers perceived increase in average annual temperature in the study area. Only 8% perceived decrease in the temperature, while the remaining percentage of respondents (3.31%) are indifferent to the changes in temperature. The trend analysis of the meteorological data of temperature between 2000 and 2014 shows an increasing trend. The regression between average annual temperature and time shows that an increase in one-year period results in an increase in the average temperature of the area by 0.003 °C (Fig. 2). Thus, farmers' perception appears to be in consonant with the statistical record of temperature from the meteorological station. This result is also in line with (Chepkoech et al. 2018), who also found increasing trend in the temperature of study area.

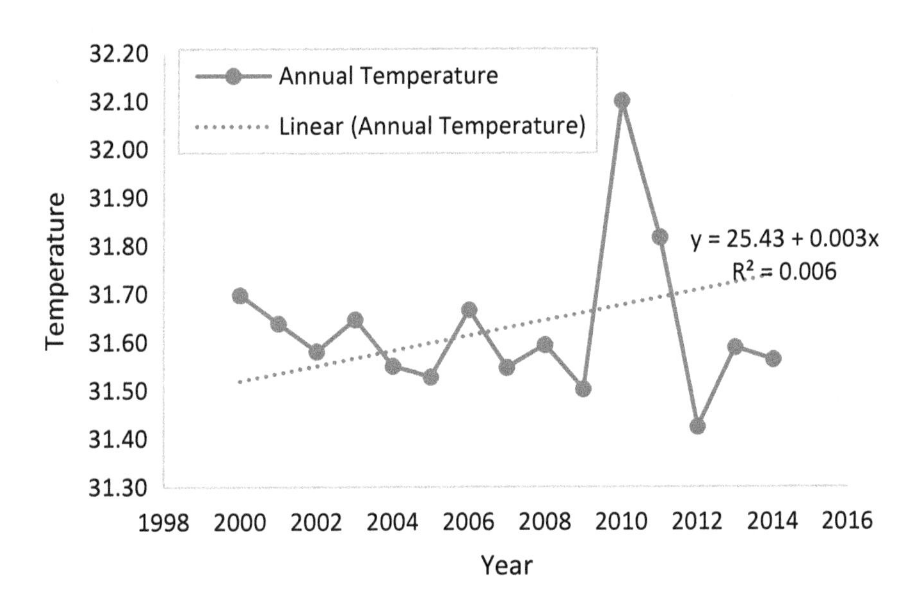

Fig. 2 Annual temperature trend in Southwest Nigeria

Farmers' Perception of Resultant Effects of Climate Change Variables

The outcome of changes in average annual rainfall and temperature results into different anomalies such as drought frequency, insects, pests and diseases infestations, loss of soil fertility, reduction in yield, among others (Chepkoech et al. 2018). Table 3 presents the perception of UIVs farmers to some of these anomalies brought about by climate change in the study area in relation to UIVs production. About 88%

Table 3 UIVs farmers' perception of resultant effects of climate change variables

UIVs farmers' perceived effects of climate change	Frequency	Percentage
Change in drought		
No change	0	0.00
Yes, increasing	161	88.29
Yes, decreasing	14	7.33
Indifference	6	3.14
No response	10	5.24
Change in pests and insects infestation		
No change	0	0.00
Yes, increasing	145	75.92
Yes, decreasing	16	8.38
Indifference	11	5.76
No response	19	9.95
Change in Soil fertility		
No change	0	0.00
Yes, increasing	28	14.66
Yes, decreasing	122	63.87
Indifference	31	16.23
No response	10	5.24
Change in UIVs Yield		
No change	0	0.00
Yes, increasing	36	18.85
Yes, decreasing	141	73.82
Indifference	4	2.09
No response	10	5.24
Change in annual Earnings		
No change	0	0.00
Yes, increasing	36	18.85
Yes, decreasing	139	72.77
Indifference	5	2.62
No response	11	5.76
Changes in the Land area allotted to UIVs		
No change	0	0.00
Yes, increasing	129	67.54
Yes, decreasing	50	26.18
Indifference	1	0.52
No response	11	5.76

Source: Field survey, 2016

of the UIVs producers believe that occurrence of drought has increased over the years as a result of reduction and/or erratic pattern of rainfall. The excerpts of the farmers during the FGDs further establish the negative effects of drought on the production activities of UIVs:

> The erratic rainfall has brought problem to farmers because most of what we plant did not germinate on time and some even got burnt in the soil as a result of prolonged drought
>
> FGD with Farmers in Ilesha, Osun State.

About 76% of the respondents indicated that the effect of insect infestation has increased greatly in recent times, compared to the past. The excerpts from the FGDs affirm this:

> There is reduction in our output due to climate change; there are some insects destroying our farm produce. To the extent that we have to take some of the species of the insect to laboratory for the scientist to help us find solution to it because the insecticide we have been using are no longer effective
>
> FGD with farmers in Ilesha, Osun State.
>
> . . . also what we have experienced this year has never been experienced before. That we plant all vegetables and insect and pest spoilt some specific vegetables. Pest also spoilt all the maize and some other crops.
>
> FGD with farmers in Ile-Ife, Osun State.

> There is no positive impact. Our expectation has been dashed because there is irregularity in rainfall. Our profits are low, insects infested our farm and destroyed it
>
> FGD with farmers in Iwaroka, Ekiti State.

Also, the response of UIVs farmers to the effect of climate change on soil fertility is negative. About 64% of the farmers are of the opinion that soil fertility has decreased greatly. More so, most (73.82%) UIVs farmers experienced reduction in yield and, as a result of this, about 73% testified to the fact that this, invariably, has drastically reduced their earnings from UIVs production. Excerpts from the FGDs conducted buttress these facts thus:

> Since there is irregularity in rainfall, we don't really get the normal output, so it has reduced our income
>
> FGD with farmers in Igboho, Oyo State.
>
> Although, there is good market for vegetables now, but the insect infestation has reduced our output which resulted in low income, and this had brought financial difficulties on farmers
>
> FGD with farmers in Ilesha, Osun State.

Comparison of Responses of UIVs to Changes in Rainfall

An understanding of the responses of UIVs to the indirect effect of climate change variables such as yield reduction, insect, pests, and diseases infestations is very crucial. This will guide farmers on the type of climate change adaptation strategies to adopt in order to reduce the negative impact of climate change on UIVs production. The responses of the farmers across the four States are harmonized and summarized in Table 4. Unanimously, the farmers across the four States agreed that the UIVs' response to resultant effects of climate change differs and it is premised on the type of UIVs and season. This finding is in an agreement with Chepkoech et al. (2018) who also found that the sensitivity of African Indigenous Vegetables to climate change variables is determined by the season and the type of vegetable, for instance, in the case of amaranth species, fluted pumpkin, jute mallow, and African eggplant; extreme rainfall does not really have effect on both their yield and the incident of insects and pests' infestations, although they experience moderate disease infestation such as yellowing of leaves. During the dry season, their yields are moderately low and moderate rate of pests and insects' infestation occur. The disease infestations are rare during this period for the vegetables.

During the high rainfall and low rainfall periods, osun, glossy nightshade, garden huckleberry, and quail grass are moderately affected in terms of yield, pests, insects, and disease infestations. In the case of field pumpkin and fire weed, there is moderate reduction in yield and disease infestation but not affected by pests and insects during the excessive rainfall period. Extremely low rainfall or dry spell reduced the yield moderately and the effect of insects and pests too is moderate and there is no disease infestation for field pumpkin. But for fire weed, the dry spell reduced the yield greatly; in fact, the fire weed may be totally unavailable in the market. While the yield of Bologi is moderately reduced by excessive rainfall, that of waterleaf is not affected. The two UIVs are not affected by insects and pests and diseases infestation during the extreme high and low rainfall but their yields are greatly reduced during the low rainfall. It is noteworthy that lots of nutrients are washed and leached away during the excessive rain fall period. From the result, it is obvious that excessive rainfall encourages diseases infestation while low rainfall or dry spell encourages insects and pests' infestation. The result is in harmony with Chepkoech et al. (2018) results. The ranking of the UIVs that are cultivated by the farmers in the study area in order of drought tolerance is also presented in Table 4; ugu is the most drought tolerant while Bologi is ranked last on the list.

Adaptation Strategies Adopted by the UIVs Farmers

A number of adaptation strategies are adopted by the UIVs farmers in order to weather the negative effects of climate change (Fadairo et al. 2019) presented in Table 4. The adaptation strategies adopted by the UIVs farmers are presented in Table 5. From the Table, only about 15% of the farmers do not adopt any strategy. This result is similar to Mahouna et al. (2018), who also identified only 14.2% of the sample respondent as nonadopter of climate change adaptation strategy but at variance with Fadairo et al. (2019) who found that all vegetable farmers adopted one form of adaptation strategy or the other. From the discussion with the farmers, it was gathered that the most important adaptation strategies adopted by UIVs farmers

Table 4 Comparison of responses of UIVs to resultant effects of changes in rainfall

	Response of UIVs to very high rainfall				Response of UIVs to very low rainfall				
UIVs	Yield	Insects and pests infestation	Diseases infestation	Soil fertility	Yield	Insects and pests infestation	Diseases infestation	Soil fertility	Order of drought tolerance
White amaranth	No change in yield	Not affected	Moderate infestation	Leached Nutrient	Moderate reduction in yield	Moderate infestation	Not affected	Moderate effect	5
Red amaranth	No change in yield	Not affected	Moderate infestation	Leached Nutrient	Moderate reduction in yield	Moderate infestation	Not affected	Moderate effect	5
Fluted pumpkin	No change in yield	Not affected	Moderate infestation	Leached Nutrient	Moderate reduction, tough leaves	Moderate infestation	Not affected	Moderate effect	1
Field pumpkin	Moderate reduction in yield	Not affected	Moderate infestation	Leached Nutrient	Moderate reduction in yield	Moderate infestation	Not affected	Moderate effect	4
Jute mallow	No change in yield	Not affected	Moderate infestation	Leached Nutrient	Moderate reduction, tough leaves	Moderate infestation	Not affected	Moderate effect	3
African eggplant	No change in yield	Not affected	Moderate infestation	Leached Nutrient	Moderate reduction in yield	Moderate infestation	Not affected	Moderate effect	2
Osun	Moderate reduction in yield	Moderate infestation of pests	Moderate infestation	Leached Nutrient	Moderate reduction in yield	Moderate infestation	Moderate infestation	Moderate effect	4

(continued)

Table 4 (continued)

UIVs	Response of UIVs to very high rainfall				Response of UIVs to very low rainfall				Order of drought tolerance
	Yield	Insects and pests infestation	Diseases infestation	Soil fertility	Yield	Insects and pests infestation	Diseases infestation	Soil fertility	
Glossy nightshade	Moderate reduction in yield	Moderate infestation of pests	Moderate infestation	Leached Nutrient	Moderate reduction in yield	Moderate infestation	Moderate infestation	Moderate effect	4
Garden huckleberry	Moderate reduction in yield	Moderate infestation of pests	Moderate infestation	Leached Nutrient	Moderate reduction in yield	Moderate infestation	Moderate infestation	Moderate effect	4
Quail grass	Moderate reduction in yield	Moderate infestation of pests	Moderate infestation	Leached Nutrient	Moderate reduction in yield	Moderate infestation	Moderate infestation	Moderate effect	6
Fire weed	Moderate reduction in yield	Not affected	Moderate infestation	Leached Nutrient	High reduction in yield	–	–	–	7
Bologi	Moderate reduction in yield	Not affected	No infestation	Leached Nutrient	High reduction in yield	No infestation	No effect	Moderate effect	9
Waterleaf	Not affected	Not affected	No infestation	Leached Nutrient	High reduction in yield	No infestation	No effect	Moderate effect	8

Table 5 Adaptation strategies adopted by the UIVs farmers

Adaptation strategy	Frequency	Percentage	Position
No adaptation	29	15.18	8th
Crop diversification	51	26.70	3rd
Changing time of planting	45	23.56	4th
Diversification to other nonagricultural activities	121	63.35	2nd
Agroforestry and perennial plantation	7	3.66	9th
Use of integrated pest management and use of fertilizer	39	20.42	7th
Cultivating along river banks	187	97.91	1st
Good practices (mixed cropping, crop rotation, mulching, organic fertilizer)	44	23.04	5th
Irrigation	41	21.47	6th

over the years is to cultivate UIVs along the river bank during the dry season and use upland during the wet season to reduce the incident of diseases and flood from excessive high rainfall. About 98% adopt this strategy. The information gathered further reveals that about 63% of UIVs producers are diversifying from agriculture to nonagricultural related businesses as a result of their unpalatable experiences in farming. Some of the excerpts from the FGDs show that many farmers are discouraged as a result of some of their experiences. The third popular adaptation strategy option is crop diversification. About 27% of the respondents indicated crop diversification as adaptation strategy they have adopted. An excerpt from the FGD in Ile-Ife, Osun State, shows this:

> ... that was why I changed to okra plantation. When I tried it and it performed better, I decided to change to okra cultivation
>
> FGD with farmers in Ile-Ife, Osun State

Also, 23.56% changed time of planting, and about 23% also adopted agricultural good practices such as mulching, crop rotation, and mixed cropping. About 21% adopted irrigation, but the excerpts from the FGD show that where rain failed, some adopted irrigation system and it led to outbreak of insect infestations which pesticide could not handle. The excerpt from the FGD in Osun State reveals this thus:

> The first step we took when the rain did not fall was to use irrigation. Different pests and insect infestations showed up. We used different insecticides to kill the insects, but chemical compositions were no longer effective like before.
>
> FGD with farmers in Ilesha, Osun State

It was noteworthy that the least (3.66%) adaptation strategy adopted by the UIVs farmers is agroforestry and perennial plantation, and about 15% do not adopt any adaptation strategy.

Variable Description of Adopters and Nonadopters of UIVs Producers

The description of the socioeconomic characteristics of UIVs farmers and other relevant variables to this study are summarized in Table 6. The information provided in the table shows that the nonadopters are significantly older than the adopters. There is no significant difference between their years of formal education, UIVs vegetable land area, access to climate change information, agro-ecological zone, and average monthly temperature of adopters and nonadopters in the study area. Meanwhile, the adopters had significantly higher mean values of the years of UIVs production experience, net revenue from UIVs, off-farm income, average monthly precipitation, farm distances from market and main road than the nonadopters.

Table 6 Variables and summary statistics

	Nonadopters		Adopters		
Variable	Mean	SD	Mean	SD	Mean difference
Age of respondent	45.28	13.65	42.18	14.16	3.10***
Years of formal education	9.07	4.80	9.23	4.68	−0.17
Years of UIVs production experience	10.79	7.24	12.79	9.61	−2.00***
UIVs Land area	0.11	0.18	0.22	0.49	−0.12
Access to climate information (1/0)	0.48	0.51	0.73	0.45	−0.25
Agro ecological zone	0.69	0.47	0.85	0.36	−0.16
Net revenue	9,420,215	14,100,000	16,700,000	27,020,000	−7,276,625***
Off farm income	5,47,848.30	5,80,313.8	1,043,479	2,109,412	−4,95,630.40***
Average monthly temperature	31.44	2.33	31.66	2.36	−0.22
Average monthly precipitation	114.95	91.98	116.54	82.39	−1.59***
Farm distance from market	5.60	4.02	6.12	4.57	−0.52***
Farm distance from main road	1.70	1.58	2.21	3.41	−0.53***

*** = Significant at 1% level

Determinants of the Decision of UIVs Farmers to Adopt Climate Change Adaptation Strategies

Binary logistic regression analysis was used to analyze the determinants of the decision to adopt climate change adaptation strategies by the UIVs farmers in the study area. Variables such as revenue from UIVs, age, sex, household size, education, marital status of the respondents, vegetable farm size, farm distance to market and main road, experience in vegetable farming, access to climate information, climate change awareness, and agroecological zone are included in the analysis. The results of the analysis reveal that the revenue generated from the UIVs business has a positive and significant influence at the 1% threshold on the adoption of climate change adaptation strategy. The more the revenue generated from the business, the more likely the producer will be willing to adopt adaptation strategy. This implies that adoption of adaptation strategy is not free and resources for the implementation must be available for farmers to adopt. The age of UIVs producers is significant at 10% level but negative. The implication of this is that as farmers increase in age the likelihood of adopting adaptation strategy reduces. This finding is in line with Uddin et al. (2014), that as farmers age, they lose interest in adopting climate change adaptation strategy. The number of years of experience in UIVs farming has positive and significant relation at 10% threshold on the adoption of adaptation strategy. This is expected, and more experience implies more competence in weather forecasting. This outcome is similar to the result of Mahouna et al. (2018) who also found that the more the experience of farmers the more the likelihood of adopting climate change adaptation strategies. Access to climate information is also significant and has positive influence at the 5% threshold on adoption of adaptation strategy. This implies that the more access farmers have to sources of information on climate change, the more likely they will adopt adaptation strategy to ease the negative effect of climate change on their business. More so, awareness of climate change is significant at 1% level and has positive effect on the likelihood of farmers to adopt adaptation strategy. This means that when farmers are aware and well informed about climate, the likelihood to adopt adaptation strategy will not be difficult. Agro ecological zone too is significant and positively determines the likelihood of farmers adopting adaptation strategy. This suggests that the effect of climate change in various agro ecological zones differ and as such, the need for climate change adaptation strategy may also differ. Lastly, access to loan is significant at 10% level and positively relates to choice of adopting adaptation strategy. Those farmers who have access to loan will likely adopt adaptation strategy since the strategies are not free (Table 7).

Conclusion and Recommendations

This chapter presents a microlevel study on the perceived effect of climate change and adaptation strategy employed on UIVs business in Southwest Nigeria. To this end, data set from plot level survey of 191 UIVs farm was used in the analysis. UIVs

Table 7 Determinants of the decision of UIVs farmers to adopt adaptation strategy

Adoption	Odds ratio	Standard error	z-value
Revenue	38.9202	55.3609	2.57***
Age of respondents	−0.9555	0.065	−1.64*
Sex of respondent	1.9052	1.2175	1.01
Household size	0.9895	0.0559	−0.19
Years of formal education	0.9603	0.0656	−0.59
Marital status	2.0791	2.3835	0.64
Experience in UIVs production	1.0680	0.0416	1.69*
Access to climate information	3.5169	1.9898	2.22**
Vegetable land area	3.3851	6.0988	0.68
Climate change awareness	46.7125	39.6536	4.53***
Agro ecological zone	3.0296	1.8863	1.78*
Farm distance to main road	1.0304	0.1398	0.22
Farm distance to market	0.9825	0.0710	−0.24
Membership of association	0.9301	0.5787	−0.12
Access to loan	4.4184	3.9790	1.65*
Constant	−9.97e-11	8.30e-10	−2.77***
Logistic regression	Number of observation = 191 Wald chi2 (15) = 63.84 Prob>chi2 = 0.0000 Pseudo R2 = 0.4018 Log likelihood = −47.5208		

*, **, *** = Significant at 10%, 5%, 1% level, respectively

production activities, adaptation strategies information, and meteorological data were also obtained. In conclusion, 13 economically viable UIVs are identified across the study area and their responses to the indirect climate change effect such as insects, pests, and diseases infestation, soil fertility, drought and yield, differ. While some UIVs are not affected by these climate change effects, some are moderately affected and to some others, the effect was great. The study also concludes that the occurrence of insects and pests' infestations is common whenever rain cease to fall for a long time while diseases infestation is common during the excessive rainfall in the study area during the raining season. The study further concludes that UIVs farmers adopt nine different adaptation strategies to ameliorate the effect of climate change variables on the UIVs business. The most prominent of them is the cultivating UIVs along river banks and the least practiced is agroforestry and perennial plantation.

Finally, the study concludes that the factors that determine the likelihood of farmers to adopt climate change adaptation strategy are revenue from UIVs, age of UIVs farmer, years of experience in UIVs production, access to climate information, climate change awareness, agro ecological zone, and access to credit. From the conclusion, the following recommendations are made:

- In the era of climate change, where many crops are failing, promotion of production, marketing, and consumption of UIVs is advocated for.
- Development of improved varieties of UIVs that are more tolerant to increased rainfall is needed.
- Information on climate change should be made available and accessible to farmers.
- Farmers should be trained on the right adaptation strategy to adopt, considering the type of crop they cultivate and the peculiarity of their agro-ecological zone.

Acknowledgments This work was supported by the UK Department for International Development under the Climate Impact Research Capacity and Leadership Enhancement (CIRCLE) Visiting Fellowship programme. Neither the findings nor the views expressed, however, necessarily reflect the policies of the UK Government.

References

Abewoy D (2018) Review on impacts of climate change on vegetable production and its management practices. Adv Crop Sci Tech 6(330):7. https://doi.org/10.4172/2329-8863.1000330

Actionaid International (2011) On the Brink: who's prepared for a climate and hunger crisis. Retrieved from http://www.actionaid.org/publications/brink-whos-best-prepared-climate-and-hunger-crisis. 2011, in Third World News

Adebooye OC, Opabode JT (2004) Status of conservation of the indigenous leaf vegetables and fruits of Africa, 3(December), pp 700–705

Aju P, Labode P, Uwalaka R, Iwuanyanwu U (2013) The marketing potentials of indigenous leafy vegetables in southeastern Nigeria. Int J AgriSci 3(9):667–677. Retrieved from www.inacj.com

Ayanwale A, Oyedele D, Adebooye O, Adeyemo V (2011) A socio-economic analysis of the marketing chain for under-utilised indigenous vegetables in Southwestern Nigeria. In: African crop science conference proceedings, pp 515–519

Ayyogari K, Sidhya P, Pandit MK (2014) Impact of climate change on vegetable cultivation – a review. Int J Agric Environ Biotechnol 7(1):145–155. https://doi.org/10.5958/j.2230-732X.7.1.020

Boko M, Niang I, Nyong A, Vogel C, Githeko A, Medany M (2007) Africa. In: Parry ML, Canziani OF, Palutikof JP, van der Linden PJ, Hanson CE (eds) Climate change 2007: impacts, adaptation, vulnerability. Contribution of working group II to the fourth assessment report of the intergovernmental panel on climate change. Cambridge University Press, Cambridge, UK

Chepkoech W, Mungai NW, Stöber S, Bett HK, Lotze-Campen H (2018) Farmers' perspectives: impact of climate change on African indigenous vegetable production in Kenya. Int J Clim Change Strat Manag 10(4):551–579. https://doi.org/10.1108/IJCCSM-07-2017-0160

Ching LL, Stabinsky D (2011) Ecological agriculture is climate resilient, Third World Network – TWN Durban Briefing Paper (No. 1). Retrieved from www.twnside.org.sg

Chivenge P, Mabhaudhi T, Modi AT, Mafongoya P (2015) The potential role of neglected and underutilised crop species as future crops under water scarce conditions in sub-Saharan Africa. Int J Environ Res Public Health 12(6):5685–5711. https://doi.org/10.3390/ijerph120605685

Devi AP, Singh MS, Das SP, Kabiraj J (2017) Effect of climate change on vegetable production – a review. Int J Curr Microbiol App Sci 6(10):447–483. https://doi.org/10.20546/ijcmas.2017.610.058

Ebert AW (2014) Potential of underutilized traditional vegetables and legume crops to contribute to food and nutritional security, income and more sustainable production systems. Sustainability (Switzerland) 6(1):319–335. https://doi.org/10.3390/su6010319

Enete IC (2014) Impacts of climate change on agricultural production in Enugu state, Nigeria. J Earth Sci Clim Change 5(9):3. https://doi.org/10.4172/2157-7617.1000234

Eyzaguirre P, Padulosi S, Hodgkin T (1999) Priority setting for underutilized and neglected plant species of the Mediterranean region. In: Padulosi S (ed) IPGRI's strategy for neglected and underutilized species and the human dimension of agrobiodiversity. International Plant Genetic Resources Institute, Allepo, pp 1–20

Fadairo O, Williams PA, Nalwanga FS (2019) Perceived livelihood impacts and adaptation of vegetable farmers to climate variability and change in selected sites from Ghana, Uganda and Nigeria. Environ Dev Sustain. https://doi.org/10.1007/s10668-019-00514-1

Guarino L (1997) Traditional Africa vegetables: promoting the conservation and use of underutilized and neglected crops. In: Proceedings of the IPGRI International Workshop on Genetic Resources of Traditional Vegetables in Africa, Rome

Howden M, Soussana JF, Tubiello FN (2007) Adaptation strategies for climate change. Proc Natl Acad Sci 104:19691–19698

Intergovernmental Panel on Climate Change (2014) Climate change 2014 synthesis report summary chapter for policymakers. IPCC

IPGRI (2002) Neglected and underutilized plant species: strategic action plan of the International Plant Genetic Resources Institute, Rome

Kothari CR (2004) Research methodology, methods and techniques (second rev). New age International (P) Limited, Publisher, New Delhi

Lemmi LK (2013) Climate change perception and smallholder farmers' adaptation strategy: the case of Tole District, Southwest Showa Zone, Oromiya Regional State, Ehiopia. Department of Rural Development and Agricultural Extension, Haramaya University

Mabhaudhi T, O'Reilly P, Walker S, Mwale S (2016) Opportunities for underutilised crops in southern Africa's post–2015 development agenda. Sustainability 8(4):302. https://doi.org/10.3390/su8040302

Mahouna A, Fadina R, Barjolle D (2018) Farmers' adaptation strategies to climate change and their implications in the Zou Department of South Benin. Environments 5(15):17. https://doi.org/10.3390/environments5010015

Maroyi A (2011) Potential role of traditional vegetables in household food security: a case study from Zimbabwe. Afr J Agric Res 6(26):5720–5728. https://doi.org/10.5897/AJAR11.335

Nnamani C, Oselebe H, Agbatutu A (2009) Assessment of nutritional values of underutilized indigenous leafy vegetables of Ebonyi State, Nigeria. Afr J Biotechnol 8(9):2321–2324

Omatseye TR (2009, September 28) Speech by the Director General, NIMASA, at the 2009 World Maritime Day Celebrations, Abuja, Nigeria. In: Iwori J (ed) GLOBAL WARMING: poverty to worsen in Nigeria. THISDAY Newspaper, p 31

Padulosi S, Hoeschle-Zeledon I (2004) Underutilized plant species: what are they? Leisa Margazine, 5

Padulosi S, Heywood V, Hunter D, Jarvis A (2011) Underutilized species and climate change: current status and outlook in crop adaptation to climate change. (eds: Yadav SS, Redden RJ, Hatfield JL, Lotze-Campen H, Hall AE). Blackwell, Oxford

Raghuvanshi RS, Singh R (2001) Nutritional composition of uncommon foods and their role in meeting in micronutrient needs. Int J Food Sci Nutr 32:331–335

Sambo BE (2014) Endangered, neglected, indigenous resilient crops: a potential against climate change impact for sustainable crop productivity and food security. J Agric Vet Sci 7(2):34–41

Shrestha D (2013) Indigenous vegetables of Nepal for biodiversity and food security. Int J Biodivers Conserv 5(March):98–108. https://doi.org/10.5897/IJBC11.124

Tanton T, Haq N (2008) Climate change: an exciting challenge for new and underutilized crops. In: Smartt J, Haq N (eds) New crops and uses: their role in a rapidly changing world. Centre for Underutilized Crops, University of Southampton, Southampton

Uddin MN, Bokelmann W, Entsminger JS (2014) Factors affecting farmers' adaptation strategies to environmental degradation and climate change effects: a farm level study in Bangladesh. Climate 2:223–241. https://doi.org/10.3390/cli2040223

10

Case for Climate Smart Agriculture in Addressing the Threat of Climate Change

John Saviour Yaw Eleblu, Eugene Tenkorang Darko and Eric Yirenkyi Danquah

Contents

Abstract

Climate smart agriculture (CSA) embodies a blend of innovations, practices, systems, and investment programmes that are used to mitigate against the adverse effects of climate change and variability on agriculture for sustained food

J. S. Y. Eleblu (✉) · E. Y. Danquah
West Africa Centre for Crop Improvement, University of Ghana, College of Basic and Applied Sciences, Accra, Ghana

E. T. Darko
Geography and Resource Development, University of Ghana, Legon, Ghana

production. Food crop production under various climate change scenarios requires the use of improved technologies that are called climate smart agriculture to ensure increased productivity under adverse conditions of increased global temperatures, frequent and more intense storms, floods and drought stresses. This chapter summarizes available information on climate change and climate smart agriculture technologies. It is important to evaluate each climate change scenario and provide technologies that farmers, research scientists, and policy drivers can use to create the desired climate smart agriculture given the array of tools and resources available.

Keywords

Climate change · Climate · Climate smart agriculture · Food security · Breeding approaches

Introduction

Background

Climate describes the weather conditions of a region such as its temperature (hot, warm, or cold) which is due to amounts or intensity of sunshine, rainfall (dry or wetness) and its pattern, air pressure, humidity, cloudiness, and wind, throughout the year, averaged over a series of years. "Climate change" as a terminology was suggested by the World Meteorological Organization (WMO) in 1966 to represent climate variations over long periods of time often from decades to millennia, irrespective of the causative agents (Hulme 2017). The term has been widely accepted and has fast become a household name for climatic variations which are often not favorable for man's survival. Climate change has largely been associated with anthropogenic global warming; however, it is indeed larger and encompasses all vagaries in climatic conditions which occur over decades. Also human activities are estimated to have caused approximately 1.0 °C of global warming above pre-industrial levels, with a likely range of 0.8 °C to 1.2 °C. Global warming is likely to reach 1.5 °C between 2030 and 2052 if it continues to increase at the current rate (IPCC 2018). In today's world, the term climate change has evolved from being a technical jargon for describing vagaries in climatic pattern into a global issue agent which requires the intervention of man to prevent future disastrous outcomes being predicted.

It should be noted that this book chapter will cover very limited information on climate change as the objective is to guide the reader to appreciate the need for a response that adopts innovations to accelerate the development of climate smart agriculture technologies as mitigation efforts against climate change. With that understanding we shall proceed to attempt to cover the breadth of knowledge in a summary of what is known with regards to the expected impact of climate change on crop production and food security, an overview of climate smart agriculture technologies and what is possible given current trends in technology and innovation.

Even though mitigation and adaptation responses compete with each other due to potential negative trade-offs across spatial, temporal, institution (Smith and Olesen

2010), economic scales (Wilbanks and Sathaye 2007). While mitigation measures aim to reduce emissions on a global scale, adaptive measures are specific to micro-environments and address various local impacts of climate change. As a result of the interconnection between the environment and socio-economic risks, the agriculture sector offers opportunities for complementary actions through the implementation of ecosystem sensitive approaches known as the CSA. This new approach is to bridge the growing divide between the two discourses and foster long-term resilient development in the agriculture sector. CSA is defined by FAO as "agriculture that sustainably increases productivity, resilience (adaptation), reduces/removes GHG's (mitigation) and enhances achievement of national food security and development goals' (FAO 2010). Therefore, adaptation, mitigation, and food security are the three key pillars of CSA (Lipper et al. 2014).

Climate smart agriculture (CSA) is a way to achieve sustainable development as well as green economy goals. It intends towards food availability and takes part to conserve natural assets and is closely associated with perception of improved growth, as FAO develops it for crop yield (FAO 2011). There is a high need for climate smart agriculture because agricultural production systems are expected to produce food for a global population of about 9.1 billion people in 2050 and over 10 billion by the end of the century (UNFPA 2011). This, however, has necessitated the development of CSA strategies and policies at different levels of governance (Zougmore et al. 2016). Therefore, it is highly imperative to sustain livelihoods which are predominantly agrarian in these regions.

Climate Change and Food Security

Climate change has the potential to threaten food production and, consequently, food security especially in vulnerable regions. One major area where the impact of climate change is expected to be very significant in threatening the very existence of humanity is the estimated effect of climate change on agriculture. Agriculture is the major source of income and livelihood for an estimated 70% of the poor and vulnerable people who live in rural areas with limited resources oftentimes without access to basic technologies (World Bank 2016b). However, the production of food is being affected by climate change, it is therefore important to study the influence of this global climate change to meet the requirements of people and is estimated that by 2100, the world population will reach about 10 billion (Boogaard et al. 2014). The climate change and variability will adversely impact on food security and agriculture livelihoods of the poorest farmers, fisher folks and forest dwellers. Even though sub Saharan Africa contributes less than 5% of the global greenhouse gas (GHG) emissions, the region is vulnerable to the negative effects of climate change due to the fact that the region's development prospects are closely linked to the climate due to the great reliance on rainfall (Tol 2018). Added to other non-climatic stresses (poverty, inequality, and market shocks), the impact of climate change will make negatively impede the achievement of the Sustainable Development Goals (SDGs) on livelihoods, food security, poverty reduction, health, and access to clean water in vulnerable communities (IPCC 2014a). However, the use of climate change predictions based on theories and past data accrued over centuries is difficult to use in projecting the expected impact of changing climates on food security. Since the institution of climate change as a body or a field of study, many academic and scientific publications have emerged. The first scientific work by Katz, published in the first issue of the first journal on climate change titled "Climate Change" on the effect of climate change on food production clearly questioned the accuracy of any such predictions and warned that the predicted impacts at the time were estimates should be acknowledged as such. A direct quote from Katz follows: *"Attempts to assess the impact of a hypothetical climatic change on food production have relied on the use of statistical models which predict crop yields using various climatic variables. It is emphasized that the coefficients of these models are not* universal constants, but rather statistical estimates subject to several sources of error. Thus, any statement regarding the estimated impact of climatic change on food production must be qualified appropriately" (Katz 1977).

The aforementioned challenges have been addressed by leading investigators recently where climate change impact has been modelled based on quick country-level measurement of vulnerability to food insecurity under a range of climate change and adaptation investment scenarios (Richardson et al. 2018). The findings have been made accessible through their publication and an online interactive portal that is user friendly for policymakers (Met Office 2015). The interactive graphically displayed model predicts that food insecurity vulnerability is anticipated to worsen rapidly under all simulations of GHG emissions, and the re-distribution of vulnerable geographic regions remains very similar to present-day conditions where sub-Saharan Africa and South Asia remain the most severely affected. By the year 2050, an additional 2.4 billion people expected to be living in developing countries with much concentration in South Asia and sub-Saharan African, where agriculture is an important sector and major employment source, but currently more than 20% of the population is on average food insecure (Wheeler and von Braun 2013). About 75% of the global poor live in rural areas, and agriculture is their most important source of income (International Fund for Agricultural Development 2011). High levels of adaptation is seen to be able to decrease vulnerability across affected areas; however, the only scenario with the highest level of mitigation combined with high levels of adaptation shows appreciable levels of reduction in vulnerability compared to the present-day prevailing conditions (Smith and Olesen 2010). As agriculture is directly affected by climate change, adaptation strategies are becoming increasingly important issues for promoting development (Clements et al. 2011). Therefore, adaptation strategies in the context of climate change are all those practices that are employed by smallholder farmers to either get used to or minimize the effects of climate change and variability. According to the IPCC, adaptation is the process of adjustment to actual or expected climate and its effects that in human systems, adaptation seeks to moderate or avoid harm or exploit beneficial opportunities (IPCC 2014). The strategies for adapting to climate change and variability can be grouped into two; namely, autonomous and planned adaptation strategies. The autonomous adaptation strategies involve actions taken by non-state agencies such as farmers, communities, or organizations and/or firms in response to climatic shocks while planned adaptation involves actions taken by local, regional, and or national government to provide infrastructure and institutions to reduce the negative impact of climate change. However, the planned adaptation which measures or results from deliberate policy decisions and awareness from farm to global levels and are discussed in literature as key to reducing present and future vulnerability and climate impacts on livelihoods (IPCC 2014). However, there are limitations to planned adaptation measures under severe conditions. As a result, more systematic changes in adaptive capacity and resource allocation are being considered. So in this discussion we shall look at the various climate smart agriculture practices that can help mitigate the climate change effects on agriculture. Therefore, the effects of climate change can be solved by climate smart agriculture practices such as climate smart crop (breeding), improved pasture and animal rearing, amelioration of degraded lands, rehabilitation of polluted water bodies, and management of sustainable systems such as agroforestry, livestock management, and manure management. Also, the promotion of sustainable land management practices which are also part of CSA practices (Branca et al. 2011) have influenced paradigms shift from the traditional practices. Most of these technologies can help mitigate greenhouse gasses (GHG) emissions. Food security and improving food productivity can also reduce human vulnerability to climate impacts and the need for additional land conversion to agriculture, which represents almost as much as GHG emissions and those directly generated from agriculture activities (IPCC 2014), but food production and security measures may conflict with climate smart and conservation objectives, especially intensifying agriculture and producing more food for a growing population (Matocha et al. 2012).

Climate Smart Agriculture Technologies

The climate smart agriculture technologies will focus on describing some of the approaches which include breeding (climate smart crop), efficient resource management, integrated renewable energy technologies for farming systems, resource conserving technologies, land use management, cropping season variation, efficient pest management, forecasting, and geographic information system (GIS) mapping.

Breeding and Climate Change

Agriculture was born about 13,000 years ago when man gradually transitioned from hunting and gathering lifestyle into domestication of wild plants and animals. Food production systems since the invention of Agriculture which remains heavily dependent on the availability of rainfall has been evolving progressively to match-up to the growing demands of the human population. It is noteworthy that the art of breeding which emerged through domestication which involved selecting plants and animals that were acceptable with good qualities for the consumption/utilization by man. Breeding which begun as selection has seen many advancements; notable are hybridization techniques, matting designs and schemes, genetics enhanced hybridization programs, tissue culture and mutagenesis aided systems, genetic engineering using recombinant DNA technologies, and Genomics- and other Omics-assisted breeding and the latest being genome editing. A summary of the resources and tools available for breeders in this day and age is presented in Box 1 below. As breeding has evolved based on man's knowledge and the development of tools to aid the development of new variants, climate and the rate of climate change has rapidly outpaced and outstripped the worlds production systems especially in areas of greatest vulnerability where new technologies remain inaccessible. The dry areas and flood prone areas are of the greatest concern where extreme weather conditions can prevail and persist for long periods disrupting the natural seasons and cycles of production that farmers are used to. These concerns can mostly be addressed if all tools available to breeders are widely accepted and utilized to aid in the development of climate smart crops that are designed to adapt to harsh and extreme weather conditions producing higher yields compared to currently available varieties that do poorly under such conditions.

Box 1 Array of Tools and Resources Available to Breeders

Genetic Resources
Gene Banks
Mutants
Core Collections
Diverse Panels
Bi-Parental QTL
Recombinant Inbred Lines
Nested Association Mapping Population
MAGIC
Training Populations

Genetic Resource Characterization
Microscopy
Basic Phenotyping
Genotype-by Environment Studies
Screen houses & Greenhouses
Hydroponics
Aeroponics
Environment Simulation
Live imaging
Advanced phenotyping platforms
Satellite-Aided Phenotyping

1st Generation Breeding Tools
Domestication/ Selection
Hybridization techniques
Vegetative Propagation Techniques

2nd Generation Breeding Tools
In vitro propagation techniques
Organogenesis and Embryo rescue
Anther culture
Somaclonal variations
In situ conservation
In vivo dissection and analysis

3rd Generation Breeding Tools
Molecular Biology Tools
QTL Mapping
Marker Assisted Breeding
Sequencing
Targeting nduced Local Lesions in Genomes

4th Generation Breeding Tools
Next Generation Sequencing
Genomics Aided Breeding
Epigenomics
Transcriptomics
Gene Expression Regulation
Metabolomics
Proteomics
Genome Editing
Comparative Genomics

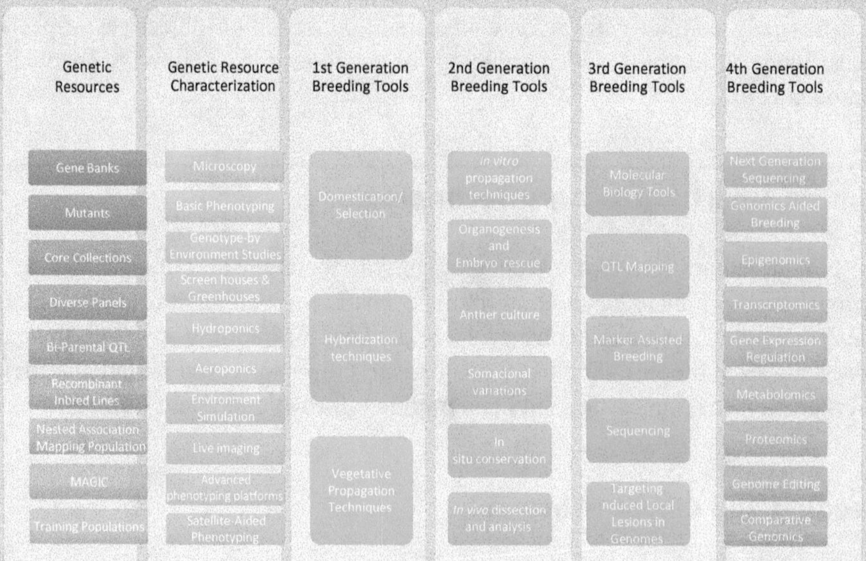

The myriad of available resources range from genetic resources available that are conserved in situ, ex situ, or in vitro; gene banks, core and representative collections not forgetting diverse panels in national, international, and regional research Centres as well as the Bi-parental, Recombinant Inbred Lines (RILs), Nested Association

Mapping, Multi-parent Advanced Generation Inter-cross (MAGIC) & Training populations in the hands of researchers and Scientists who originated and curated them. These genetic resources are sources of alleles of great agronomic importance required in the development of climate smart crops or animal breeds that can withstand and yield highly under changing climatic conditions. These are therefore the first range of arsenals of breeders in the fight against dwindling productivity under climate change conditions. The next in the array of tools are those that can be used to characterize, evaluate, detect, select, and then release these climate adaptable varieties to farmers and businesses for increasing productivity under prevailing circumstances. These tools have evolved from first generation to the current cutting edge fourth generation tools that are available for breeders to use in the development of new and improved varieties with better adaptation to the changing environmental conditions. The first-generation tools mainly encompass discoveries of basic principles of domestication/selection, the knowledge and use of pollination to make self and crosses as well as means of vegetative propagation such as grafting, corms, bulbs etc. Second generation array of tools are mainly based on advances in biology that allow for cell, tissue and organ culture which allow for more advanced technologies in the crop and animal improvement by breeders. Third and fourth generation tools such as represented in Box 1 that have been developed add speed and precision to the array of tools that are currently available for quick development of improved climate smart crops.

It is important to evaluate each climate change scenario and provide strategies breeders can use to create the desired climate smart crops given the array of tools and resources available. The various climate change scenarios, potential impact and climate smart breeding approaches are delineated in Table 1. For instance, climate change is greatly impacting agriculture currently in the tropics and other arid areas with erratic rainfalls that no longer follow the patterns or established seasons known to farmers that heavily depend on rain-fed crop production systems. This scenario has the potential impact of poor yields as a consequence of untimely start of the farming activities and crop failure. For adaptability to such scenarios, climate smart crops with adaptability to different types of droughts or erratic rains could be developed using a combination of tools and resources described in Box 1 and made available to farmers. Such climate smart crops are required in vulnerable areas threatened by climate change in order to avert the worsening food insecurity problem and ensure the achievement of sustainable development goals 1 (no poverty) and 2 (no hunger).

Efficient Resource Management

Another approach that can be of relevance in achieving the objectives of climate smart agriculture is efficient management of resources. This approach is an important part of CSA and the future environment. In the food production chain, from the farmer to the customer or final consumer, almost one third part of the food is lost due

Table 1 Climate change scenarios, potential impact, and climate smart breeding approaches

Climate change scenarios	Potential impact	Climate smart approaches
Erratic rainfall	Farmers plant too late or too early leading to yield losses	Climate smart crops with adaptability to different types of droughts or erratic rains.
Prolonged droughts in arid areas and New Droughts prone areas	Crop losses, Famine and loss of lives	Drought resistant crops that perform well under water limited conditions
Increased floods	Damage to crops and animals; loss of lives and property, displacement of people	Development of water loving crops as well as crops resistant to lodging
Intense rain and wind storms	Damage to crops and animals	None
Rise in sea levels	Increase in salt stress on crops, loss of arable lands to toxic levels of salts, low or no yields	Salt resistant or tolerant crops
Loss of soil cover	Soil erosion, loss of soil fertility, loss of microbes in the soil	Planting of trees and plants that will rehabilitate the soils. Introduction of bioengineered microbes that encourage soil health.
Increased global temperatures	Reduced yields, new pests and disease emergence and damage to crops and animals	Improved heat adaptable crops

to the improper management of resources (Hartter et al. 2017). On yearly basis, for instance, the total energy consumption in the global food losses are almost 38% of all the energies utilized by the food chain. Critical areas in the food chain processes which serve as good avenues for improving energy efficiency includes: transportation, conservation, processing, cooking, and consumption (FAO 2011). In Africa, a majority of wood removed is used for manufacturing household articles as well as cooking. However, cooking in stoves helps save energy thereby decreasing deforestation in the long run. For instance, this technique of managing resources and climate change projects has helped in supporting sustainable intensification through a number of initiatives including the establishment of an agriculture information and the decision support system and the preparation of soil management plans. Since this approach was adopted in 2014, climate smart agriculture was adopted on 2,946,000 hectares and has provided for a carbon sequestration potential of up to 9 million tons carbon dioxide annually, (https://www.worldbank.org/en/topic/climate-smart-agriculture). Additionally pastoralists are also enjoying some benefits from climate smart agriculture in the Sahel, Burkina Faso, Chad, Mali, Mauritania, Senegal, and Niger. The application of rangeland management is boosting productivity and resilience. This approach is also helping to reduce emissions.

Integrated Renewable Energy Technologies of Farming Systems

The Integrated Renewable Energy Technologies is the application of suitable energy technologies, tools, and different farming services which are relevant in creating the stable change to energy smart proficient food systems. These technologies are governed by conditions of nature. These technologies are very useful because in the long run there will be a reduction in GHG's emissions. For instance, mid-season aeration can be promoted through short term drainage. Some of the technologies in the energy smart food system are the windmills, solar panels, wind generators, photovoltaic lights, biogas, and conversion of hydrothermal tools, bio energy and water pumping machines, information and communication technological innovation, and other similar approaches (Bochtis et al. 2014; Basche 2015). This technology has been applied in Morocco through the Morocco inclusive project Green Growth project, through the supply of agrometeorological information and the facilitation of the resilience building technology such as direct seeders. The pumps used can be both fuel and electric water pumps which are mostly used on irrigation farming (deep well and submersible pumps). Stakeholders in the agricultural industry should appreciate this modern technological innovation due to the benefits of increasing the value of production in the farming business. Most times, these technologies are linked on the farm from an integrated food energy system as shown in Fig. 1 below.

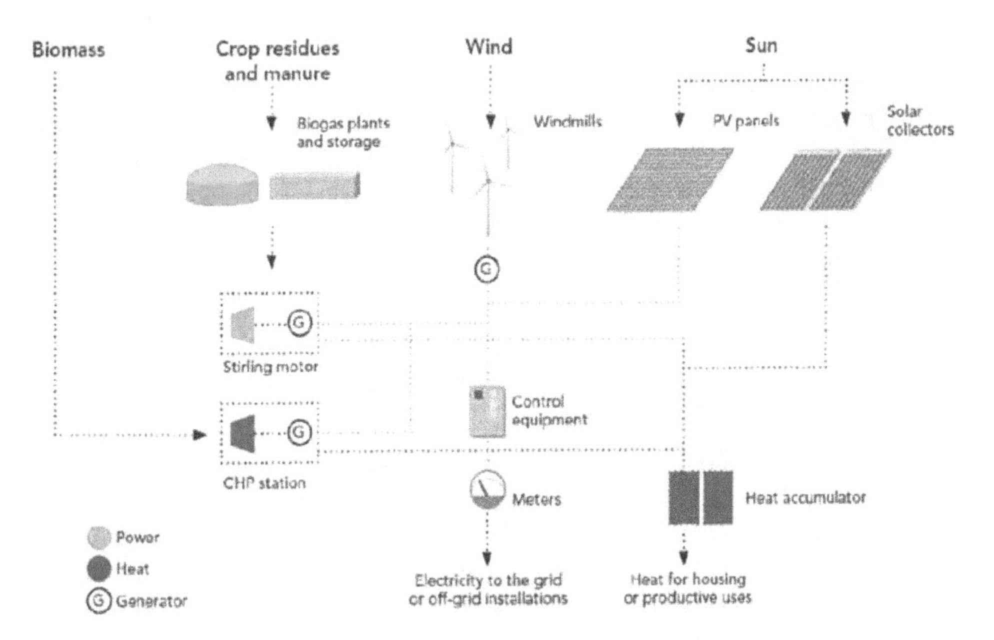

Fig. 1 Integrated renewable energy technologies for farming systems. (Source: Amin et al. 2016)

Resource Conserving Technologies (RCTs)

This technology consists of methods that enhance efficiency in the management of inputs. When these resource technologies are implemented, it comes with its own merits which includes low cost of productivity, limited use of fuel, labor, water, and early planting of crops which results in improved yields in the long run. For instance, the zero tillage system, which happens to be one of the resource conserving technologies, is a type of cultivation system in which the seeds are directly sowed into a virgin (uncultivated) soil. The zero-tillage system, however, involves the cultivation of crops into untilled soil by aeration of thin channels with adequate depth-width so as to attain suitable seed coverage. The soil remaining is left as if tillage has never been done on it before (Derpsch et al. 2010). In some parts of the globe, the zero tillage permits farmers to grow wheat very soon after the rice harvest. This allows the head of the crops to appear and the filling of the grains before the warm weather, pre-monsoon set off. Therefore, as the average temperature of the globe in certain parts rises, early planting will be more relevant for the production of wheat (Pathak 2009).

Land Use Management

Land use management involves the proper planning of land and its usage. The proper planning includes fixing the location of plants and livestock production, changing the concentration of the application of plant foods and bug sprays can reduce global warming on agricultural activities (Ahmad et al. 2014). Other land use practices involve shifting production out of marginal areas, changing the role of applying cartilage and pesticides. However, it must be noted that capital and labor can minimize the risks from Climate change on agriculture production. The farmers can regulate the duration of the growing season by changing the time at which the farming fields are sown. Other adaptation mechanisms can be in the form of changing the times of irrigation and use of fertilizer.

Figure 2 elucidates Cropland Expansion Potential for different continents.

Cropping Season Variation

The planting dates can be set to reduce the infertility that is caused by increasing temperatures; this may prevent the flowering period from meeting with the hot period (Arslan et al. 2015). The effects of the increased climatic variations which normally happens in both the semiarid and arid regions sometimes take advantage of the wet period by changing the planting times/dates. However, this approach is usually to avoid intense weather events in the growing season. This system of cultivation promotes the development of strong cultivars thereby leading to the production of different crops. The planting dates can be set to reduce the infertility that is caused by increasing temperature; this may prevent the flowering period from

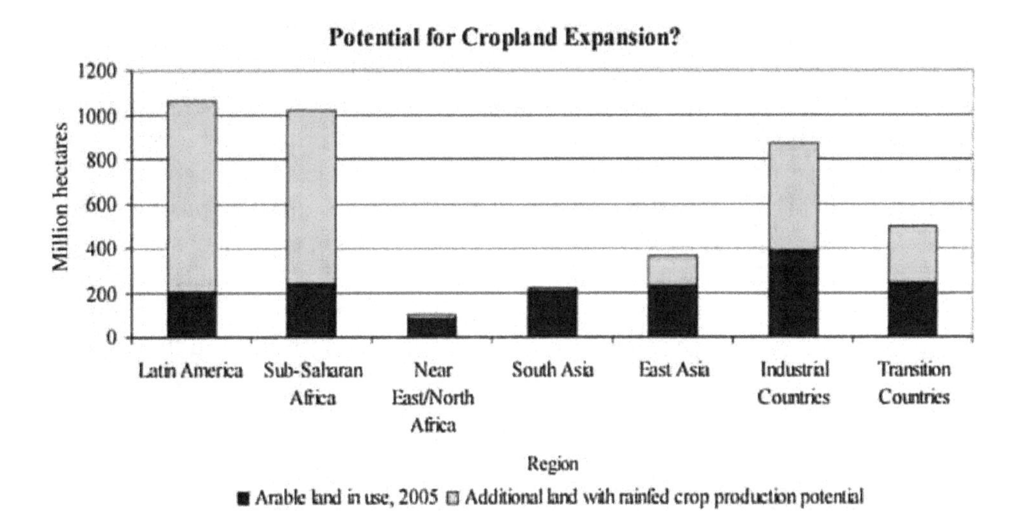

Fig. 2 Land use management. (Source: Burnisma 2009)

meeting with the hot period. The effects of the increased climatic variations which normally happens in both the semiarid and arid regions sometimes take advantage of the wet period by changing the planting times/dates. However, this approach is usually to avoid intense weather events in the growing season. This system of cultivation promotes the development of strong cultivars thereby leading to the production of different crops. The farmers will, therefore, need to ensure that they adopt the changing crop rotation system in the various hydrological cycles (Pathak et al. 2012).

Crop Relocation

This approach involves the grouping and sorting of the plants and the varieties with respect to its sensitivity to the weather condition of a place. Crop relocation helps the crops to perform well according to the sensitivity of the climate during the vegetative and productive stages (Shames et al. 2012). There are several factors which affect agricultural production as a result of climatic change. These include increase in temperature, carbon dioxide (Co_2) levels, and increase in drought and floods. These impacts vary across the various regions in the world as well as the different cultures. Other factors such as daylight, temperature, and humidity are very necessary for the vegetative and reproductive growth of the crops. Additionally the period for harvesting should be properly done so as to minimize losses during the period (Baba et al. 2017). However, it is therefore important to differentiate regions and crops that are highly susceptible to climate change. For instance, it is obvious that temperature increases affect the quality of many important crops. Some of these crops with respect to the discussion include basmati rice and tea.

Efficient Pest Management

The normal agriculture pest and insect management on farms poses a threat to the environment. The usage of chemicals to destroy and kill pests is very harmful for farmers and also living organisms in the soil. Even chemicals such as insecticide, herbicides for plant diseases have been banned by some governments of certain countries, as the situation exists now, there is no botanical or environmental friendly chemical available. Due to this, farmers still use the chemicals for controlling pests on their farms. So, basically this technique provides an opportunity to employ environmentally friendly measures for pest control. The difference in the climatic factors such as the fall and rise in temperature unpredictably influences pest and disease incidence thereby impacting on major crops.

Therefore, the change in climate will affect the relationship of pest and weed and the host populations. However, some of the adaptation strategies in this pest management approach includes;

(i) Improvement in different breeding types that are resistant to pest and disease.
(ii) Strong pest adaptation mechanism with more relevant control for both biological and cultural practices.
(iii) Adoption of techniques such as crop substitution with regards to places resistant to pest and hazards.

GIS Mapping

This approach is used in analysis and mapping. It is a system which is designed to capture, store, manipulate, analyze, manage, and present geographical data.

Fig. 3 Risk Assessment and mapping. (Source: Amin et al. 2016)

However, it helps in the estimation and computation of storm causes and flooding that is related to hot cyclones. The study in using GIS mapping considers factors such as property allocation, infrastructure facilities among other resources. The photograph and images (Fig. 3) were used in the experimentation of seashore due to rising sea levels and hot cyclones. The figure below shows risk which can be explained by the cumulative study of emerging threat and the existing patterns of vulnerability. The technique enables the creation of hazards and risk maps at many different possible scales or dimensions to show the threat allocation across different geographical spaces within the globe. Some of the geographical places can be site specific, municipal (administrative areas) and other natural landscapes in river basins, coastlines, and lakes. Figure 3 portrays Mapping and Risk Assessment.

Conclusion

Climate change is a great threat to agriculture and as such there is the need to tackle this adverse impact by adopting new innovative techniques in climate smart agriculture. This chapter has dealt with some of the climate smart agriculture techniques that can help reduce the impacts of climate change on agriculture and increase food crop production. To achieve food security and agriculture development goals, adaptation to climate change will be required to lower emission intensities per output. Thus improving food protection by moderate climate change, sustainable use of the natural resources, using all products more competently, have less inconsistency and greater constancy in their outputs. More fruitful and more flexible agriculture requires a paramount change in the usage of resources such as land, water, soil nutrients, and genetic resources management by climate smart agriculture approaches.

References

Ahmad MI, Ali MA, Khan SR, Hassan SW, Javed MM (2014) Use of crop growth models in agriculture: a review. Sci Int 26(1):331–334, Lahore

Amin A, Mubeen MHM, Nasim W (2016) Climate Smart agriculture: an approach for sustainable food security. Agriculture Research Communications 3:13–21

Arslan A, Mccarthy N, Lipper L, Asfaw S, Cattaneo A, Kokwe M (2015) Climate smart agriculture? Assessing the adaptation implications in Zambia. J Agric Econ 66(3):753–780

Baba RA, Owiyo T, Barbie B, Denton F, Rutabingwa F, Kiema A (2017) Advancing climate-smart agriculture in developing drylands: joint analysis of adoption of multiple on-farm soil and water conservation technologies in West African Sahel. Land Use Policy 61(2):196–207

Basche AD (2015) Climate –smart agriculture in Midwest cropping systems: evaluating the benefits and trade-offs of cover crops. PhD Dissertation, Iowa State University

Bochtis DD, Sorensen CGC, Busato P (2014) Advances in agricultural machinery management: a review. Biosyst Eng 126(10):669–681

Boogaard H, De Wit A, Te Roller J, Van Diepen C (2014) Wofost control centre; users guide for the Wofost control centre and the crop growth simulation model Wofost 7. Alterra, Wageningen

Branca G, Mccarthy N, Lipper L, Jolejole MC (2011) Climate smart agriculture: a synthesis of empirical evidence of food security and mitigation benefits from improved cropland management. Retrieved from http://www.fao.org/climatechange/297640aa5796a4fb093b6cfdf05558.pdf.

Burnisma J (2009) The resource outlook to 2050: By how much to land, water use and crop yields need to increase by 2050. Expert meeting on how to feed the world in 2050. FAO and ESDD, Rome, p 33

Clements R, Haggar J, Quezada A, Torres J (2011) In: Zhu X (ed) Technologies for climate change adaptation-adaptation sector. UNEP Riso Centre, Roskilde

Derpsch R, Friedrich T, Kassam A, Li H (2010) Current status of adoption of no till farming in the world and some of its main benefits. Int J Agric Biol Eng 3:1–25

FAO (2010) Climate- smart agriculture: policies, practices and financing for food security, adaptation and mitigation. Food and Agriculture Organization of the United Nations, Rome

FAO (2011) The state of food and agriculture 2010–11. In: Women in agriculture: closing the gender-gap for development. Food and Agriculture Organisation of the United Nations, Rome

Hartter J, Hamilton LC, Boag AE, Stevens FR, Ducey MJ, Christo ND, Palace MW (2017) Does it matter if people think climate change is human caused? Clim Serv 10(4):53

Hulme M (2017) Climate change, concept of. Int Encycl Geogr People Earth Environ Technol 16. https://doi.org/10.1002/9781118786352.wbieg0343

IPCC (2014a) Climate change: impacts, adaptation and vulnerability. Part A: Global and sectoral aspects. Contribution of working group II to the fifth assessment report for the intergovernmental panel on climate change. Cambridge University Press, Cambridge, UK/New York

IPCC (2014b) Chapter 7: Food security and food production systems. In: Climate change 2014: impacts, adaptation, and vulnerability. IPCC working group II contribution to AR5. IPCC Secretariat, World Meteorological Organization, Geneva. Available: http://ipccwg2.gov/images/uploads/WGIIARS5-Cha

IPCC (2018) Summary for policymakers. In: Global warming of 1.5 °C. An IPCC special report on the impacts of global warming of 1.5 °C above pre-industrial levels and related global greenhouse gas emission pathways, in the context of strengthening the global response to the threat of climate change, sustainable development and efforts to eradicate poverty [V. Masson-Delmotte, P. Zhai, H. O. Pörtner, D. Roberts, J. Skea, P. R. Shukla, A. Pirani, W. Moufouma-Okia, C. Pean, R. Pidcock, S. Connors, J. B. R. Mathews, Y. Chen, X. Zhou, M. I. Gomis, E. Lonnoy, T. Maycock, M. Tignor, Waterfield (eds)]. World Meteorological Organization, Geneva, 32p

Katz RW (1977) Assessing the impact of climatic change on food production. Clim Chang 1:85–96

Lipper L, Thornton P, Campbell BM, Baedeker T, Braimoh A, Bwalya M, Torquebiau EF et al (2014) Climate- smart agriculture for food security. Nat Clim Chang 4:1068–1072. https://doi.org/10.1038/nclimate2437

Matocha J, Schroth G, Hills T, Hole D (2012) Integrating climate change adaptation and mitigation through agroforestry and ecosystem conservation. In: Nair PKR, Garrity D (eds) Agroforestry-the future of global land use. Springer, Dordrecht, pp 105–126

Met Office (2015) Food insecurity and climate change model spread. Retrieved from http://www.metoffice.gov.uk/food-insecurity-index

Pathak H (2009) Agriculture and environment. In: Handbook of agriculture. Directorate of Information and Publication Agriculture, ICAR, New Delhi, pp 62–92

Pathak H, Aggarwal PK, Singh SD (eds) (2012) Climate change impact, adaptation and mitigation in agriculture: methodology for assessment and applications. Indian Agricultural Research Institute, New Delhi, p 302

Richardson KJ et al (2018) Food security outcomes under a changing climate: impacts of mitigation and adaptation on vulnerability to food insecurity. Clim Chang. https://doi.org/10.1007/s10584-0182137-y

Rural Poverty Report (2011) International fund for agriculture development

Shames S, Wollenberg E, Buck LE, Kristjanson P, Masiga M, Biryahaho B (2012) Insitutional innovations in Africa smallholder carbon projects. CCAFS Report No. 8. Copenhagen, CCAFS

Smith P, Olesen J (2010) Synergies between the mitigation of, and adaptation to climate change in agriculture. J Agric Sci 148:54–552. https://doi.org/10.1017/S0021859610000341
Tol RSJ (2018) The economic impacts of climate change. Rev Environ Econ Policy 12:4–25
UNFPA (2011) Annual Report: Delivering Results in a World of 7 billion, New York
WB (2016a) World Data Bank –agriculture indicators. The World Bank, Washington, DC. Assessed 05 June 2020
WB (2016b) World Data Bank-agriculture indicators. The World Bank, Washington, DC. Accessed 05 June 2018
Wheeler T, von Braun J (2013) Climate change impacts on global food security. Science 341:508–513
Wilbanks T, Sathaye J (2007) Integrating mitigation and adaptations responses to climate change: a synthesis. Mitig Adapt Strateg Glob Chang 12(5):957962. https://doi.org/10.1007/s110270079108*3
Zougmore R, Partey S, Ouedraogo M, Omitoyin B, Thomas T et al (2016) Toward climate-smart agriculture in West Africa:a review of climate change impacts, adaptation strategies and policy developments for the livestock, fishery and crop production sectors. Agriculture and Food Security 5(1):26

11

Equity and Justice in Climate Change Adaptation: Policy and Practical Implication in Nigeria

Chinwe Philomina Oramah and Odd Einar Olsen

Contents

Abstract

Over the past decade, justice and equity have become a quasi-universal answer to problems of environmental governance. The principles of justice and equity emerged as a useful entry point in global governance to explore the responsibilities, distribution, and procedures required for just climate change adaptation. These principles are designed primarily through the establishment of funding mechanisms, top-down guides, and frameworks for adaptation, and other adaptation instruments from the UNFCCC process, to ensure effective adaptation

C. P. Oramah (✉) · O. E. Olsen
Department of Safety, Economics, and Planning, University of Stavanger, Stavanger, Norway
e-mail: chinwe.p.oramah@uis.no; oddeinar.olsen@uis.no

for vulnerable countries like Nigeria that have contributed least to the issue of climate change but lack adaptive capacity. Global adaptation instruments have been acknowledged for adaptation in Nigeria. Climate change has a detrimental impact on Nigeria as a nation, with the burden falling disproportionately on the local government areas. As Nigeria develop national plans and policies to adapt to the consequences of climate change, these plans will have significant consequences for local government areas where adaptation practices occur. Although the local government's adaptation burden raises the prospects for justice and equity, its policy and practical implication remains less explored. This chapter explores the principles of justice and equity in national adaptation policy and adaptation practices in eight local government areas in southeast Nigeria. The chapter argues that some factors make it challenging to achieve equity and justice in local adaptation practices. With the use of a qualitative approach (interview (n = 52), observation, and document analysis), this chapter identified some of the factors that constraints equity and justice in local government adaptation in southeast Nigeria.

Keywords

Adaptation policy · adaptation practices · environmental justice · equity · Nigeria · local level

Introduction

The gap between the developed and developing nations as regards development is extending to the risks and security issue of climate change. While almost all countries are affected by the risk and security impacts of climate change, it is widely recognized that developing countries are more vulnerable, lacks adaptive capacity (Tabbo and Amadou 2017), and would suffer disproportionately (Stallworthy 2009; Rübbelke 2011). This is particularly the case for developing countries in Africa living in poverty (IPCC 2014). Wide recognition that industrialized countries are overwhelmingly responsible for climate change has slowly led to conceptualizing adaptation as a global issue with a formidable dilemma of equity and justice in developing countries such as Nigeria (McManus et al. 2014; Thomas and Twyman 2005). Scaling adaptation as a global issue recognizes an international responsibility to provide financial support and funding for undertaking adaptation at the national or local level in developing countries (Benzie and Persson 2019; Ciplet et al. 2013; Saraswat and Kumar 2016). Despite the funds obtained by some developing countries for climate change adaptation, implementing adaptation at the local level continues to be challenging.

Many of the impacts of climate change, such as floods and drought, are experienced at the local level (Rauken et al. 2015). As a result, the burden of climate change adaptation practices falls disproportionately on the local government area. Adaptation is a localized phenomenon that addresses local circumstances with the need for local solutions and actions (Corfee-Morlot et al. 2011; Measham et al. 2011; Moore 2012; Nalau et al. 2015). Conceptualizing climate change adaptation as a local phenomenon is based on the principle of subsidiarity, which is a belief that tasks should be trusted with the lowest level, where the local actors are always able and willing to govern their natural resources effectively (Lockwood et al. 2009). This approach to adaptation assumes that local actors have the required resources to practice adaptation in isolation.

However, local government in Nigeria is embedded in a broader multiscale governance context comprising a range of government actors from the state, the federal, and global levels. Thus, current thinking poses that adaptation plans should be understood and developed at national and subnational levels, practiced at the local level, and funded via international institutions (Benzie and Persson 2019). As a result, local adaptation is increasingly supporting and driving adaptation initiatives and policies within the framework provided by national and state-level legislation (Vogel and Henstra 2015). The extent to which these adaptation policies consider vulnerabilities and impacts of climate change at the local level as well as the extent to which local government participates in instituting national adaptation policies and frameworks are debatable. Against this backdrop, different scholarly voices have emerged over the years, arguing that adaptation decision-making at policy and practical levels has justice and equity implications (Few et al. 2007; Paavola and Adger 2006; Thomas and Twyman 2005).

Nigeria has instituted national adaptation plans and policies as well as established climate change institutions to aid adaptation. Despite the adaptation policies and institutions, some local government areas struggle to cope with and respond to climatic impacts. This is because there are no regulations and institutions designed to foster climate change adaptation in the local government area (Oulu 2015). Presently, the local government institutions carrying out adaptation, such as the ministry of environment and planning, department of works, and local emergency management agency (LEMA), are not designed for climate change challenges. This chapter discusses prospects for justice and equity principles in Nigeria's national adaptation policy and local adaptation practices. This chapter affirms that equity and justice are the important normative goal in both national and local climate change adaptation, but argue that the practice of equity and justice in climate change adaptation is often embedded in the illusion of inclusion. Without due consideration of equity and justice at all levels of governance, there would be a tension between the underlying principle of fair adaptation and participation by the local governments and vulnerable groups in Nigeria. Alternatively, a more instrumental approach to appropriate adaptation at the local level is more likely to succeed as long as local government inclusion is made explicit from the outset.

This chapter discusses the perceived role of equity and justice regarding national adaptation policy at the federal level and adaptation practices in eight local government areas in southeast Nigeria. With the use of a qualitative approach through interview and document analysis, this chapter explores the principles of equity and justice in policies and practices of climate change adaptation in Nigeria.

Conceptualizing Justice and Equity as It Relates to Climate Change Adaptation

Climate change issues give concern for different types of justice: distributive, procedural, recognition, compensatory, and restitutive justice (Ciplet and Roberts 2017; Khan et al. 2019; Klinsky and Dowlatabadi 2009; Rawls 1971). The basic structure in the subject of climate change justice here is that different communities experience differentiated impacts of climate variability in part by the political system as well as by economic and social circumstances. This differentiated physical and social vulnerability to climate change impacts create deep inequalities between developing and developed countries. Vulnerable developing countries lack the tools and adaptive capacity required to develop the appropriate response to climate risks. The development of tools and adaptive capacity both at the local and national level has been a significant focus on global adaptation, especially regarding equity and justice. However, equity and justice issues of adaptation are more readily discussed at the global, regional, and national levels than at the local level (Thomas and Twyman 2005), even though adaptation practices are undertaken at the local level.

At the global level, the differentiated vulnerability to the impact of climate issues was brought to the international community's attention in November 2006 at the United Nations Framework Convention on Climate Change (UNFCCC) held in Nairobi, intending to identify situations that increase or reduce the capacity to adapt (Vogel et al. 2007). It was then argued that adaptation would promote benefits that can lead to equitable and sustainable development (Adger et al. 2009). The 2015 Paris Agreement includes a global goal on adaptation through reducing vulnerability to climatic impacts, reinforcing adaptive capacity, and strengthening resilience (International Summit on Climate Change held in Paris 2015). One of the commitments of developed countries under the UNFCCC is to assist developing countries to meet their adaptation cost. If more impoverished country gains access to adaptation funds through equity and justice schemes, adaptation can be improved. Global governance is considered especially relevant for Nigeria and other developing countries, as these countries are already struggling to meet climate change's security challenges (Nightingale 2017; Nath and Behera 2011). The principle of environmental justice is focused on the existence of inequity in the distribution of environmental hazards, where the environment is understood to create a condition for social justice (Schlosberg 2013). As climate change increases, environmental justice is given more broad consideration with a growing focus on sustainability and transformative politics and practice to affirm the socio-ecological unity and the interdependence of all species.

There are essential points to why equity and justice have become two crucial concepts in climate change adaptation discourse at the global and national levels. First is the principle of justice, which emerges as a reaction to the claim that devastating climate extremes such as flood, drought, and desertification made worse by climate change pose additional negative implications for vulnerable developing countries and poverty-affected communities (Nay et al. 2014). The susceptibility to climate risk goes beyond biophysical vulnerability to include human well-being, social, economic, and political factors underlying social vulnerability (Kelly and Adger 2000; Otto et al. 2017), which put vulnerable countries in a constant state of crises. Therefore, adaptation should be evaluated based on justice criteria that would benefit all groups of society as well as the future generation by providing the information and resources needed for adaptation, especially for those most vulnerable to climate change impacts.

Second, equity is a concept referring to fairness in the distribution of outcomes or distributive equity (Miller 1992). As regards climate change, Nay et al. (2014) argue that developing economies depend more on climate-sensitive activities that are more impacted by climate variability. They argue further that these developing economies also lack the political and organizational capacity to adapt to climatic impacts. Thus, the outcome of the equity principle should ensure that:

- The vulnerable are treated fairly for unduly bearing the burdens of climate change impacts
- There is an inclusive decision-making process
- There is an inclusive framework for taking and facilitating adaptation action
- There is a relationship between climate change adaptation and other factors that affect livelihoods (McManus et al. 2014)

However, within these developing countries, the social, institutional, and political structures can play an essential role in climate change adaptation. The relationships that exist between the individuals, the communities, and the state are also essential. Thus, adaptation at the local government is often enabled or hindered by other issues such as social structures, power relations, political and institutional structures, as well as the broader higher level of governance arrangements (Lawrence et al. 2015; Simonsson et al. 2011). These relationships often reaffirm the status quo and are likely to influence the issues of equity and justice in local adaptation practices. According to Eriksen et al. 2015, injustice and unfairness exist when the politically powerful actors set up institutions that advance agendas that exclude local knowledge, needs, and voices of the marginalized in adaptation decision-making. These powerful actors with authority further influence adaptation by claiming the right to legitimize or undermine different types of knowledge (Eriksen et al. 2015). Adaptation policies are often designed at the national level and may disproportionately affect vulnerable communities if they are excluded during policy design (Urwin and Jordan 2008). Understanding the local context of vulnerability through local participation in adaptation policies is essential for equitable and justifiable adaptation. This implies a process of social interaction and joint decision-making by stakeholders across governance scale in adaptation. However, from a systems perspective, one of the challenges facing such provision is associated with the complexity of social interactions involved in multilevel adaptation decision-making.

Climate Change in Nigeria

Nigeria is one of the most vulnerable countries and is highly dependent on climate-sensitive sectors. The country is located in the tropics that give her a hot tropical climate, consisting of variable rainy and dry seasons depending on location. Given the country's climatological cycle and size, there is a considerable range in total annual rainfall across Nigeria, from south to north, and in some regions from east to west. Wet and dry season prevails in the east and west, while a steppe climate with little precipitation is found in the far north. Temperature and humidity remain relatively constant throughout the year in the south, while the season varies considerably in the north (Ajayi et al. 2019). The most significant total precipitation is in the southeast along the coast around Bonny (south of Port Harcourt) and east of Calabar with annual rainfall around 4,000 millimetres (mm). The regularity of drought periods has been among the most notable aspects of Nigeria's climate in recent years, particularly in the north's drier regions (Akande et al. 2017; Haider 2019). These droughts indicate the considerable variability of climate across tropical Africa and severely affect the drier margins of agricultural zones occupied primarily by pastoral groups.

The southeast is one of the most developed regions in Nigeria, with the second-highest population density. In 2015, the southeast had a total population of 40 million. Southeast Nigeria falls within the latitude of 6′ N and 8′ N and longitude of 4′30′E and 7′30′E, describing the country's inland region. Southeast Nigeria is of the wet tropical type climate with mean annual temperatures between 21 °C and 34 ° C. The temperature is highest around March in the southeast (Iloeje 2009). The mean minimum temperature is relatively close to the coastal area, with annual rainfall exceeding 3500 mm (Njoku 2006; Nwagbara et al. 2013). In recent years, rainfall has become significantly more substantial in the southeast. In 2012, River Niger reached a record of 12.84 m above sea level. Water levels have also risen in upstream Cameroon, Mali, and Niger. These countries feed the River Niger and River Benue, which flow through Nigeria. River Niger flows through the southeast region leading to severe flooding. In 2012, flooding led to two million displacements and three hundred and sixty-three (363) deaths. In 2017, 12 states, including states in the southeast, were severely affected, leading to 200 deaths and over 600,000 displacements (Orji 2018). As climate change leads to more rainfall, floods disasters are becoming more devastating in Nigeria, especially in the southeast region. The southeastern region is also exposed to mild drought during the dry season.

Preparing for Climate Change Adaptation in Nigeria

Nigerian started showing a keen interest in climate change issues since 1994. The first national climate communication in 2003 was aimed at shedding more light on the consequences of climate change and its impact on developmental goals. With the support of development partners such as the United Nations Development Program (UNDP), the European Union (EU), United States Agency for International Development (USAID), as well as intergovernmental, regional organizations and nongovernmental agencies, several climate change adaptation strategies and policies have been designed and approved. Nigeria initiated a comprehensive planning process for adaptation by developing the National Adaptation Strategy and Plan of Action on Climate Change for Nigeria (NASPA-CCN). Prioritized adaptation measures in the NASPA-CCN report tend to focus on agriculture, forestry, water resources, human health, human settlement, energy, transportation and communication, industry, disaster and security, livelihoods, vulnerable groups, and education. In NASPA-CCN report, there is a recognition that climate change adaptation can best be achieved through multilevel effort requiring global, national, state, local government, nongovernmental, and civil society coordination (BNRCC 2011). In addition to this, Nigeria has instituted policies and established climate change institutions to aid adaptation. Policies such as National Policy on Erosion and Flood Control, National Water Policy, Nigeria Drought Preparedness Plan, National Forest Policy, National Health Policy, the National Policy on Environment supports (for prevention and management of disasters such as floods, drought, and desertification) and Nigeria's Agricultural Policy were developed to protect agricultural land resources from drought, desert encroachment, soil erosion, and floods (BNRCC 2011). Nigeria has established a climate change framework such as the National Framework for Application of Climate Services – NFACS (to reduce communities' vulnerability by implementing the National Agricultural Resilience Framework for the agricultural sector). Nigeria has also established a climate change department under the federal ministry of environment. The country relies on NIMET (Nigerian Meteorological Agency) and NEMA (National Emergency and Management Agency) for climate-related disaster warnings, prevention, and response. At the state level, departments of climate change are functional in some states and non-functional in others. There are no known climate change departments at the local government areas; hence, exiting ministries are carrying out adaptation actions.

Exploring the Equity and Justice Perspective of Climate Change Adaptation in Nigeria

This section of the chapter takes an equity and justice perspective of adaptation policy and practice in Nigeria, which provides a useful framework for understanding the factors that promote or hinder local government adaptation. The local government areas that are the focus of the chapter are situated in the southeast zone, where the population is (a) vulnerable to climate-related floods and mild droughts, (b) lack adaptive capacity, and (c) agitating for separation from Nigeria due to poor social and political representation. Interviews, observation, and document analyses were used as the primary data sources to explore the perceived impact of equity and justice on national adaptation policies and local adaptation practices in southeast Nigeria. This chapter analyzed the national adaptation plan and other important documents. The key documents analyzed include the National Adaptation Strategy and Plan of action on Climate Change for Nigeria (NASPA-CCN) and Nigeria Intended Nationally Determined Contribution (INDC 2016). Other documents include scientific articles, policy documents, newspapers, conference speeches, and media contents (Fig. 1).

Interviews and observations were carried out between September 2017 and January 2018. The interview was conducted (n = 52) with actors working at the federal, the state, and the local government parastatals. At the national level, ten

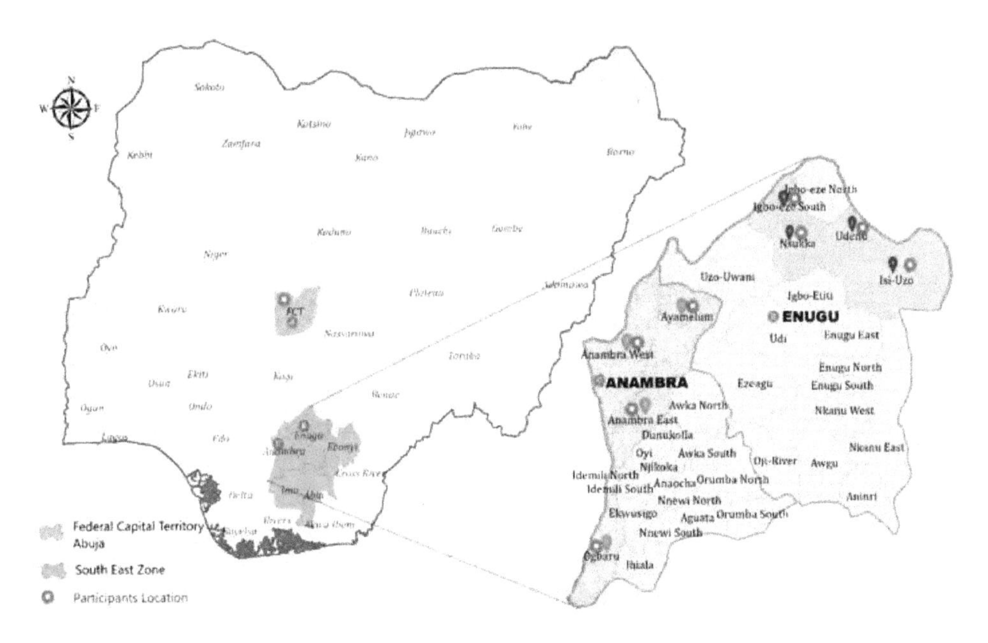

Fig. 1 Map of Nigeria showing the participant's locations

participants from the federal department of climate change were interviewed. These federal-level participants are labeled FGP (Federal government participant). At the regional level, ten participants from two states in the southeast (Anambra and Enugu state) were interviewed, where five participants were selected from each state. These participants are involved in adaptation across the state ministry of ecology, environment and climate change, the state ministry of works, and the state emergency management agency (SEMA). The state-level participants are labeled SGP (state government participants). At the local government level, 32 participants were interviewed from eight local government areas in southeast. Four experts were selected from each of the eight local government areas. These local government-level participants are labeled LGP (Local government participants). Among these participants are engineers involved in areal planning and infrastructural management, officials engaged in environmental protection, and local emergency management agencies (LEMA). Four of the local government areas, Anambra east, Anambra west, Ayamelum, and Ogbaru, are situated very close to water bodies. With the majority of the population living near the riverine area, flooding is the largest source of climate-related losses. The other four local government areas, Igbo-Eze south, Isiuzo, Nsukka, and Udenu, are situated in highlands where both floods and mild droughts are sources of climate-related losses affecting the population. Flood and drought in the southeast impact livelihood, health, crop production, livestock, groundwater dryness, and infrastructure damage. Interview questions focused primarily on how fair adaptation policies are for local government adaptation practices.

The observation was used to collect data on the adaptation practices and activities designed to cope with climate-related disasters. Data obtained through the observation method were noted and analyzed based on a follow-up question for confirmation. The interview, observation, and document analysis data were coded and categorized using Nvivo 11. The data was identified and categorized thematically using the inductive and data-oriented approach. The findings will be analyzed using the following questions as a structuring tool: what are the physiological and social vulnerability of climate change in the eight southeast local government areas in Nigeria? How is Nigeria adapting to climate change impacts? Who is responsible for adaptation policies and practices? What role do the principles of equity and justice play in adaptation in Nigeria's local government areas? These questions would be explored in three subsections. The first section will give an account of the vulnerability of local government areas to climate change. The second subsection will explain how global adaptation is affecting Nigeria's adaptation policy and practices. The last section will then explain the effect of global equity and justice on local adaptation practices.

Vulnerability and Impact of Climate Change in Nigeria

Climate change in Nigeria leads to changes in the frequency and intensity of weather and climate extremes. Nigeria's climate extremes hit people in multiple different ways. Warm temperatures cause more evaporation of water, while changes in precipitation lead to heavy rain but also swings into drought conditions. Nigeria is one of the most vulnerable countries to climate variability (IPCC 2014). The most frequently cited vulnerability is sea level rise, floods, droughts, sandstorms, landslides, erosion, intensified desertification, and general land degradation (Medugu et al. 2010). These extreme events have broad consequences for farmlands, livestock, and built infrastructures such as buildings, roads, and railways, as well as fundamental societal concerns, such as disputes over environmental resources, food security, water security, health implications, loss of livelihoods, internal and external migration, and loss of life (BNRCC 2011; IPCC 2014). The broad consequences of climate change make it imperative to assess the level of a country's vulnerability to climate change and capacity and readiness for adaptability. This chapter identified vulnerability to climate change at the national, regional, and local levels. Some vulnerability factors are frequently identified across all three scales: poverty, access to resources, livelihood opportunities, and health. Vulnerability to climate change is distributed disproportionally in Nigeria. The northeast and northwest zones are vulnerable to desertification, heat wave, loss of freshwater, intensive drought, bush burning, loss of arable lands, and livestock loss. The southeast and southwest are vulnerable to sea level rise and salinization, intensive rainfalls, floods, and damages to built infrastructures. It would be nearly impossible for preparation to be made towards adapting to these changes if the vulnerability is not adequately understood, especially from the angle of the most affected parties at the local level.

The IPCC conceptualizes vulnerability as a function of the state of a social system and the biophysical nature of climate change effects that the system face (IPCC 2007). Vulnerable to climatic impacts, Nigeria covers different frameworks.

These include risk hazards, political ecology, and socio-ecological system frameworks. Within the risk and hazard field, vulnerability is the susceptibility of people and things to losses attributable to a given level of danger, a given probability that a hazard would manifest itself in a particular way, and with a particular magnitude (Alexander 2002: 29). This field of vulnerability often neglects to address how human contribute to climatic hazards as well as the societal context in which climate hazards takes place. In political ecology, vulnerability is a characteristic of a person or group and their situation that influences their capacity to anticipate, cope with, resist, and recover from the impact of a natural hazard (Wisner et al. 2004). In the social-ecological system framework, vulnerability is a state of susceptibility to harm from exposure to stress associated with environmental and social change and the absence of capacity to adapt (Adger et al. 2006). Multiple factors such as environmental exposure, socioeconomic, political, and cultural factors operating at different levels drive vulnerability in Nigeria's local government areas. Climate hazards only lead to disaster if there is biophysical or/and social vulnerability. Biophysical vulnerability to climate change is understood as a function of environmental exposure, while the social vulnerability is a function of socioeconomic, political, and cultural characteristics of a society (Otto et al. 2017).

In Nigeria, vulnerability plays out locally. This chapter attempt an assessment of some local government areas' vulnerability to climate-related flood and drought around two dimensions, biophysical vulnerability and social vulnerability. Results show some similarities in the participant's perception of climatic impacts and adaptation measures.

Table 1 shows that flooding is the largest source of climate-related losses in four low land, local government areas in the southeast, as the participants explained. With an estimated 30% of the population living near the riverine area, in Anambra east, Anambra west, Ayamelum, and Ogbaru, loss of house settlement, loss of farmland, transportation, limited energy supply, and erosion are constant plight experienced by local communities. Out of 16 participants from low land, local government areas 14 have experienced vulnerability to flooding while 2 knows people that have experienced flood vulnerability. LG participants explained that during flooding, the only transportation system is local boats. However, they argue that using resilient structures such as iron and other metallic products to construct roads and bridges can make the situation better. Flooding causes desperate living conditions leading to the temporal fleeing of millions of people. The electric power supply in the southeast is limited at ordinary times. Flood hazards negatively influence the already limited power supply, forcing households and businesses to use generators that emit CO2 and other dangerous gasses to the detriment of human health and environmental safety.

Table 1 illustrates the frequently identified adaptation strategies used to adapt to the effect of climatic hazards. The various adaptation measures identified by LG participants are flood management, town planning, and waste management. Flood management takes place through the construction of an effective drainage system. LGP participants explained that the local communities' drainage system is weak as some drainage systems are poorly constructed. Rainwater leads to overflow and

Table 1 Local government physical and social vulnerabilities and adaptation strategies

	Local Government Area	Climate changehazards	Biophysical vulnerability to climatic hazards	Social vulnerability to climatic hazards	Effects of climatic hazards	Adaptation strategies
Lowland	1.1) Anambra east 2) Anambra west 3) Ayamelum 4) Ogbaru	Floods	Very vulnerable to exposure	1) Economic 2) Governance 3) Infrastructures	1) Loss of farmland, 2) Loss of housing 3) Erosion 4) Poor transportation 5) Energy 6) Migration	1) Flood management 2a) Avoid building on flood plains 2b) Use resilient infrastructure 3) Soil management 4) Resilient infrastructure and drainage system. 5) Energy and infrastructure management
		Mild droughts	Very low		1) Agriculture, 2) Water distribution	1) Irrigation 2) Use of deep water and boreholes
Highland	1.1) Igbo-Eze South 2) Isiuzo 3) Nsukka 4) Udenu	Floods	Moderately vulnerable to exposure	1) Lack of economic resources 2a) Dysfunctional government structure 2b) Lack of important institutions	1) Agriculture 2) Erosion 4) Transportation 4) Water security 5) Energy 6) Migration	1) Improve the agricultural system 2) Conservation agriculture and soil management 3) Use of resilient materials and functional drainage system 4) Water resource management 5) Energy and infrastructure
		Mild drought	Moderately vulnerable		1) Agriculture 2) Water distribution	1) Irrigation technology 2) Use of underground water

flood incidents because the drainage systems are often not appropriately channeled. Poor town planning leads to improper house settlements where people construct houses on floodplains. LGP participants agree that communities must adhere to town planning to avoid the loss of house settlements. Proper town planning discourages people from building on floodplains and using resilient structures that can withstand extreme weather variability. Poor waste management leads to people's disposal of waste when it is raining. These wastes block the drainage system and contribute to flooding.

Erosion is another hazard linked to climate variability in the low land, local government areas of southeast. LGP participants explain that people have lost their houses and farmlands to erosion. They also explain that erosion losses are not as severe as losses from the flood as erosion occurs slowly. LGP mentioned soil management and planting of trees as necessary measures used to reduce erosion. In Anambra east, Anambra west, and Ogbaru, the LGP participants explained that trees' planting is not sustainable due to firewood consumption.

Slow onset events such as drought are also having a substantial impact on crop production, livestock, and water distribution in low land, local government areas but at a deficient level. The use of irrigation and digging for clean water is common in these local government areas during drought. However, LGP explained that the use of irrigation is constrained by limited irrigation facilities based on available resources.

In the local government areas located in highlands, mainly: Igbo-Eze south, Isiuzo, Nsukka, and Udenu, the LGP participants explained that households are not often in danger of losing their homes due to moderate flooding instead, it is agriculture, gully erosion, road infrastructures, and energy distribution that are impacted. There is an uneven distribution of rainfall, and participants noted that the length of dry periods is on the increase. LGP participants in the local government located in highlands note that drought is a climatic hazard, leading to water shortage, with notable negative impacts on the farmers' crops, livestock, and income. These participants suggest that rainfall is often not sufficient for their agricultural production and household needs. Field observation revealed that different household sources water from streams and boreholes.

Table 1 illustrates that apart from massive flooding in the low land local government area, there are similarities in climate change challenges as well as similarities in adaptation measures in both the low and high land local government areas. The most remarkable difference between the low land local government areas and the high land local government areas is their preparedness. LGP explained that even though climate change is leading to unpredictable rainy seasons, those in riverine areas are often more prepared, which inspires more proactive adaptation strategies. In the Anam community, people come together to construct tall buildings to adapt to floods. Households are often not caught off guard as they proactively get temporal housing settlements and their boats ready for transportation as soon as flood starts.

In all the local government areas, it was indicated that the adoption of the adaptation measures frequently mentioned is moderate due to lack of resources. Though communities in the southeast are already implementing numerous strategies to cope with climate change, LG participants agree that the adaptation measures are carried out on individual, group, and community levels. The reasons given for low adaptive capacity vary from access to funds, lack of climate change awareness, and lack of human resources in the face of climate change. Nigeria context throws light into how physical vulnerability interacts with social vulnerability in climate change adaptation issues. Looking at climate change as both biophysical and social problems allows political and socioeconomic measures to evaluate the effects on the poor and vulnerable.

Adaptation Policies and Practices in Nigeria

Policy Implications of Adaptation Governance in Nigeria

Since UNFCCC first conference in 1995, nations have convened to institute and implement binding climate agreements, either as regards to mitigation or adaptation. These binding agreements have lasting impacts on how global climate treaties and national climate policies evolve. These agreements also help determine how financial resources to adapt are distributed (Gurwitt et al. 2017). Nigeria has been engaging in international climate policy negotiations since 1994 when the country becomes a party to the United Nations Framework Convention to Climate Change (UNFCCC). Nigeria ratified Kyoto Protocol in 2004 and submitted the first national climate communication in 2003 and the second national communication in 2014. United Nations Framework Convention on Climate Change provides funding to developing countries with National Adaptation Plans of Action (NAPAs). Under such treaties, countries are required to develop NAPA to adapt to climate change. Nigeria prepared its National Climate Change Action Plan in 2011, which led to the Nigeria Climate Change Policy's approval in 2012.

NAPA provides Nigeria and other least developed countries (LDCs) with an opportunity to meet their urgent and immediate needs for adapting to climate change. In 2015, Nigeria prepared its Intended Nationally Determined Contribution (INDC) and signed the Paris Agreement in 2017. These policy documents' common objective is to demonstrate political commitment to adaptation and communicate the overall government approach to adaptation. Nigeria's policy plan helps identify climate change impacts and vulnerabilities and identify areas where the country's adaptive capacity can be improved (INDC 2016). Adaptation policy targets different sectors of Nigeria's society such as agriculture, freshwater, coastal resources, forest, biodiversity, health and sanitation, human settlement, energy, transportation and communication, industry and commerce, disaster, migration and security, livelihood, education, and vulnerable groups. Nigeria Climate Change Action Plan and INDC report recognize that achieving an adaptation goal would require international support due to its low adaptive capacity. As explained in (BNRCC 2011), Nigeria National Adaptation Strategy and Plan of Action on Climate Change, Nigeria seeks to:

- Detail financial needs assessment to accurately determine the economic costs of climate change adaptation
- Revise the National Fiscal Policy to incorporate the cost of climate change adaptation
- Create a national financing mechanism to support real adaptation needs
- Access necessary international adaptation funding and technologies and manage those funds well

The above are top-down measures that would trickle down to the local level. Within the Nigerian climate policy document, the role of the federal government, the state government, the local government, the private sector, and civil society are made explicit. The federal government is responsible for instituting policies while the local government is responsible for implementing adaptation policies. The issues emphasized in Nigeria's Plan of Action are issues of collaboration, transparency, and finance. BNRCC (2011) report indicates that the federal and state governments would collaborate with the local government to strengthen communities' adaptive capacity by providing:

- Information and technological know-how, facilitating financial and other measures
- Put in place adaptation communication to allow all stakeholders to participate actively in climate change adaptation (NASPA-CCN, 2011)

However, FG participant notes that Nigeria's adaptation policy is increasingly influenced by intergovernmental organizations, as the submission of these documents and reports is relevant to obtain proper support. FG and SG participants explained that adaptation policies encompass climate change issues affecting all Nigerians and strategies to solve those issues. On the other hand, LG participants indicate that the adaptation policy is a one-fit document that lacks knowledge of local problems and solutions. Proposals for an international climate change adaptation policy recognize local representation, even though there are hardly any inquiries to ensure local representation inclusion.

Actors Perspective on Adaptation Governance in Nigeria

FG, SG, and LG participants were asked about their role in climate change adaptation practices. All the FG participants claim that they have engaged in different adaptation practices. Twenty percent of SG participants admitted that they had taken no action but are aware of several adaptation projects. The LG participants claim to have engaged in different adaptation activities such as road construction, house construction, helping community members during rescue operations, and helping to deliver aids.

When asked about who is responsible for climate change adaptation practices, FG and SG participants were quick to point fingers to developed nations. They claim that

Nigeria has benefited from international climate change adaptation funds; however, LG participants explain that individuals and groups carry out adaptation practices on a low scale, as the funds have not translated into effective adaptation practices in the local government areas in the southeast. All participants perceived the role of local government in different ways. The FG participants thought that the local government is getting the necessary resources to help communities address climate change issues. The SG participants are aware of the local government plight as it relates financial and technological resources but insists that the local government is in the position to help local communities adapt.

Interestingly, LG participants think that local government programs to address climate change are indigent. The main reason for this perception appeared to be communication and governance issues. Communication issues bother on perceived lack of consultation and transparency on the part of the state and federal government. The reason for the governance issue included the perception of autonomy and mistrust of the state and federal government. LG participants expressed that federal and state government interferes in local government matters.

When asked about the collaboration in climate change adaptation, the FG participants rate collaboration between the three government levels as excellent. The SG participants rate the relationship between the federal and state government levels as good and state with local government as fair. The LG participants thought there is almost zero collaboration between the local government and other government levels. Eighty percent of LG participants describe the collaboration between the local and the other government levels as servant–master collaboration.

Equity and Justice in Adaptation Policies and Practices in Nigeria

Global equity and justice are essential to plan and mobilize the resources needed to implement adaptation actions. However, it could not be straightforward for international policy to lay claim in sovereign affairs taking place within a sovereign territory. The dilemma of equity and justice in climate change adaptation takes different dimensions in Nigeria. In Nigeria, like many developing countries, contributing minimally to climate change issues, climate variability has become a significant threat to survival and sustainable development, especially for vulnerable individuals and communities (Ilevbare 2019). Nigeria is vulnerable to climate security issues with low adaptive capacity.

On the one hand, there is a top-down international rule system to promote adaptation ambition and accountability. On the other hand, climate change implicates domestic sensitivities in Nigeria. There is a diverging perception of how global equity and justice scheme is impacting Nigeria adaptation policies and practices. The result indicates that there is a fundamental difference between interpretations of equity and justice by FG, SG, and LG participants.

In this section, a perceived overview of adaptation policy and adaptation practices will be presented. A shared perspective on equity and justice is essential not only for transparency but also for ensuring that fair and just adaptation reaches the most vulnerable people. Those at the federal and state level view adaptation as the responsibility of the developed country that has contributed most to climate change issues, while those at the local level think adaptation is the responsibility of the national and state government. This view is reflected in the policy report, which indicates that Nigeria needs assistance from international, regional, and non-governmental organizations to reach its intended adaptation goals (INDC 2016). LG participants indicate that the vulnerable local government is struggling to meet adaptation requirements despite the fund Nigeria government acquires for adaptation projects. Results show that there are several reasons why justice eludes the vulnerable communities. FG participants claim that Nigeria, as a country, still lacks the technological and financial resources despite funding from international and regional agencies. SG and LG participants agree that resources for adaptation are lacking in all government levels but argue that other factors play a significant role in poor adaptation practices. The common factors mentioned are the institutional context, social structure, power relations, and fiscal capacity for the effective management of natural resources and adaptation funds.

In Nigeria, social structure can be viewed through institutionalized relationships organized around family, religion, education, politics, media, and economy. These institutions organize the social relationship of the southeast to other regions of Nigeria. The southeast and southwest are predominantly Christians, while the northeast and northwest are predominantly Muslims. The different zones with various ethno cultural groups merged into one country in 1914. The different zones have different tribal groups, languages, and cultures. The differences in culture, politics, and tribal identification affect people's relationships with one another. Culture and ethnoreligious politics influence the distribution of resources in Nigeria (Brown 2013).

It was previously found that political corruption and bad leadership affect the southeast zone (Ogundiya 2010). Southeast is a zone where an estimated 50% of the population feel that they are not part of Nigeria. This resentment can be attributed to the Biafra Civil War that killed millions of southeasterners from July 1967 to January 1970. The southeast feels marginalized, leading to some citizens advocating for fresh separation (Olajide et al. 2018). LG participants explained that due to Nigeria's political structure, the southeast lacks resources and infrastructures, which makes adaptation more difficult. Unequal policies and patterns of government structure driven by national and regional political and economic priorities benefit a particular segment of society while making others more vulnerable. Another issue in the southeast is that the service and industry sector are paid more attention at the expense of small-scale agriculture and fisheries (Nzeadibe et al. 2011), even though LG participants note that the farming communities are the most vulnerable in the southeast.

Nigeria operates federalism with an overconcentration of power at the national level (Akinsanya 1999). In this aspect, the politicians at the national level hold a more considerable amount of power to determine what happens to the vulnerable local population as regards climate change impact. Local government lacks autonomy and depends on the state and national government. This dependency leads to weak institutionalization and local government underutilization, allowing constant intervention from the state and federal government (Acheoah 2018). Adaptation policies are formulated at the state and national levels while adaptation practices take place at the local level. LG participants explained that local knowledge is often not sought during policy formulation, making it challenging to implement such policies in practice. When the state and federal government neglect the most vulnerable participation at the local government, vital communication that encourages collaboration is lost.

Interview and field observation indicate that Nigeria's institutional capacity for climate change adaptation at the federal, state, and local government level is undeveloped and weak. Oulu (2015) argues that establishing effective institutional frameworks is crucial for climate change adaptation. Even with the presence of adaptation policy and climate change department at the federal level, adaptation practices are carried out by the existing National or State Emergency Management Agency (NEMA and SEMA). These two agencies were not designed for climate change adaptation. Only two of the local government areas in the southeast have a Local Emergency Management Agency. NEMA is an existing risk management agency that takes the issue of climate change adaptation as one of its many functions. Climate change adaptation and mitigation goals are now assigned to the ministry and department of the environment. However, LG, SG, and FG participants explain that proven competencies and technological resources in the existing institutions are low.

The budgetary constraint is one of the factors inhibiting adaptation in local communities in Nigeria. However, participants from the federal, the state, and the local government areas have different explanations on how budget constraints hinder adaptation. FG and SG participants claim that budgetary constraints are because of Nigeria's poor economic condition. The LG participants attribute budget constraints to the local government's lack of financial independence. LG participants suggest that the local government also lacked autonomy that contributes to its lack of financial independence required to tackle the issue of climate change adaptation. Fieldwork observation indicates that local government relies on SEMA for relief and settlements for internally displaced people. LG participants explain that the local actors that know the communities well are often not consulted during these visits.

Conclusion

This chapter recognizes that adaptation practices in local governments in southeast Nigeria have equity and justice implications. Environmental equity and justice focus on ensuring that the most vulnerable communities and countries are not left to bear the burden alone. It was argued that climate justice should include mechanisms to ensure that most impacted at the local level have their interests considered (Thomas and Twyman 2005). However, this chapter found that vulnerability to climate change is mostly experienced at the local level, with the burden of adaptation falling disproportionately on the local government areas. Though Nigeria has developed adaptation policies that detail strategies to reduce and avoid climatic impacts, the local government is excluded from decision-making in adaptation policies, and thereby their vulnerability is often not reflected in the policy documents. This is because climate change adaptation policies and practices at the national level of governance are not open for representative dialogue, especially with the local government's participation. This chapter argues that by excluding the local government in the southeast in adaptation decision-making, the national adaptation plan and policies ignore local adaptation needs and knowledge and do not reflect local vulnerability. Beyond local participation, the interaction between the national authority and local knowledge needs to rely on fairness and accountability. Unless the most vulnerable adapt, risks associated with climate change could increase vulnerabilities, and more inequality.

Financial and technological resources remain crucial in helping the poor and vulnerable communities adapt to climate change risk and climate security issues. Access to these resources is vital for adaptation practices. The United Nations Framework Convention on Climate Change (UNFCCC) provides funding to ensure that the most vulnerable countries are not left to deal with climate change alone. The inclusion of local government in adaptation decision-making will ensure that global funding is easily translated into local practice and that adaptation resources are correctly channeled. This chapter also indicates that it is essential to understand issues such as social structure, power relations, institutional context, and budgetary constraints and how they affect local government adaptation. Thus, local government adaptation practices are not independent of preexisting sociopolitical and governance structures in developing countries like Nigeria. This chapter recommends the local government's inclusion in decision-making and formal adaptation governance to encourage partnership and transparency.

References

Acheoah OA (2018) Local government: the underutilized governance structure in Nigeria. Arts Soc Sci J 9(5):1–4

Adger WN, Paavola J, Huq S (2006) Towards justice in adaptation to climate change. In: Adger WN, Paavola J, Huq S, Mace MJ (eds) Fairness in adaptation to climate change. The MIT Press, Cambridge, MA, pp 1–19

Adger WN, Lorenzoni I, O'Brien KL (2009) Adaptation now. In: Adger WN, Lorenzoni I, O'Brien KL (eds) Adapting to climate change: thresholds, values, governance. Cambridge University Press, Cambridge/New York, pp 1–22

Ajayi JFA, Kirk-Greene AHM, Udo RK, Falola TO (2019) Nigeria. Encyclopædia Britannica. Retrieved from https://www.britannica.com/place/Nigeria/Resources-and-power

Akande A, Costa AC, Mateu J, Henriques R (2017) Geospatial analysis of extreme weather events in Nigeria (1985–2015) using self-organizing maps. Adv Meteorol 2017:8576150. https://doi.org/10.1155/2017/8576150

Akinsanya A (1999) Intergovernmental relations in Africa under the 1995 draft constitution. In: Uya E, Uchenna V (eds) Issues in the 1995 Nigerian draft constitution. IPPA, Calabar

Alexander D (2002) Principles of emergency planning and management. Terra Publishing, Harpenden, p 29

Benzie M, Persson Å (2019) Governing borderless climate risks: moving beyond the territorial framing of adaptation. Int Environ Agreements Polit Law Econ 19(4):369–393. https://doi.org/10.1007/s10784-019-09441-y

BNRCC (Building Nigeria's Response to Climate Change) (2011) National adaptation Strategy and Plan of Action on Climate Change for Nigeria (NASPA-CCN). Prepared for the Federal Ministry of Environment Special Climate Change Unit. Available at http://csdevnet.org/wp-content/uploads/NATIONAL-ADAPTATION-STRATEGY-AND-PLAN-OF-ACTION.pdf

Brown GM (2013) Nigeria political system: an analysis. Int J Humanit Soc Sci 3(10):172–179

Ciplet D, Roberts JT (2017) Climate change and the transition to neoliberal environmental governance. Glob Environ Chang 46:148–156. https://doi.org/10.1016/j.gloenvcha.2017.09.003

Ciplet D, Roberts JT, Khan M (2013) The politics of international climate adaptation funding: justice and divisions in the greenhouse. Glob Environ Polit 13(1):49–68. https://doi.org/10.1162/GLEP_a_00153

Corfee-Morlot J, Cochran I, Hallegatte S, Teasdal PJ (2011) Multilevel risk governance and urban adaptation policy. Climate Change 104(1):169–197

Eriksen SH, Nightingale AJ, Eakin H (2015) Reframing adaptation: the political nature of climate change adaptation. Glob Environ Chang 35:523–533. https://doi.org/10.1016/j.gloenvcha.2015.09.014

Few R, Brown K, Tompkins EL (2007) Public participation and climate change adaptation: avoiding the illusion of inclusion. Clim Pol 7(1):46–59. https://doi.org/10.1080/14693062.2007.9685637

Gurwitt S, Malkki K, Mitra M (2017) Global issue, developed country bias: the Paris climate conference as covered by daily print news organizations in 13 nations. Clim Chang 143(3):281–296. https://doi.org/10.1007/s10584-017-2004-2

Haider H (2019) Climate change in Nigeria: impacts and responses. K4D helpdesk report 675. Institute of Development Studies, Brighton

Ilevbare FM (2019) Investigating effects of climate change on health Risks in Nigeria (Online First), IntechOpen. https://doi.org/10.5772/intechopen.86912. Available from: https://www.intechopen.com/online-first/investigating-effects-of-climate-change-on-health-risks-in-nigeria

Iloeje NP (2009) A new geography of Nigeria, 5th ed. Longman, Nigeria

INDC (2016) Nigeria's intended nationally determined contribution. Available at http://www.4.unfccc.int/ndcregistry/PublishedDocuments/Nigeria%20First/Approved%20Nigeria's%20INDC_271115.pdf. Accessed, October 2017

IPCC (2007) Climate change 2007: impacts, adaptation and vulnerability. In: Parry ML, Canziani OF, Palutikof JP, van der Linden PJ, Hanson CE (eds) Contribution of Working Group II to the fourth assessment report of the Intergovernmental Panel on Climate Change. Cambridge University Press, Cambridge, p 976

IPCC (2014) Climate change 2014: impacts, adaptation, and vulnerability. In: Contribution for Working Group II to the fifth assessment report of the Intergovernmental Panel on Climate Change. Cambridge University Press, Cambridge, UK

Kelly PM, Adger WN (2000) Theory and practice in assessing vulnerability to climate change and facilitating adaptation. Clim Chang 47:325–352. https://doi.org/10.1023/A:1005627828199

Khan M, Robinson S-a, Weikmans R, Ciplet D, Roberts JT (2019) Twenty-five years of adaptation finance through a climate justice lens. Clim Chang. https://doi.org/10.1007/s10584-019-02563-x

Klinsky S, Dowlatabadi H (2009) Conceptualizations of justice in climate policy. Clim Pol 9(1):88–108. https://doi.org/10.3763/cpol.2008.0583b.

Lawrence J, Sullivan F, Lash A, Ide G, Cameron C, McGlinchey L (2015) Adapting to changing climate risk by local government in New Zealand: institutional practice barriers and enablers. Local Environ 20(3):298–320. https://doi.org/10.1080/13549839.2013.839643

Lockwood M, Davidson J, Curtis A, Stratford E, Griffith R (2009) Multi-level environmental governance: lessons from Australian natural resource management. Aust Geogr 40(2):169–186. https://doi.org/10.1080/00049180902964926

McManus P, Shrestha KK, Yoo D (2014) Equity and climate change: local adaptation issues and responses in the City of Lake Macquarie, Australia. Urban Clim:10, 1–10,18. https://doi.org/10.1016/j.uclim.2014.08.003
Measham T, Preston B, Smith T, Brooke C, Gorddard R, Withycombe G, Morrison C (2011) Adapting to climate change through local municipal planning: barriers and challenges. Mitig Adapt Strateg Glob Chang 16(8):889–909
Medugu IN, Majid RM, Johar F, Choji ID (2010) The role of afforestation programme in combating desertification in Nigeria. Int J Clim Chang Strateg Manag 2(1):35–47. https://doi.org/10.1108/17568691011020247
Miller D (1992) Distributive justice: what the people think. Ethics 102(3):555–593. Retrieved March 9, 2020, from http://www.jstor.org/stable/2381840
Moore FC (2012) Negotiating adaptation: norm selection and hybridization in international climate negotiations. Glob Environ Polit 12(4):30–48. https://doi.org/10.1162/GLEP_a_00138
Nalau J, Preston BL, Maloney MC (2015) Is adaptation a local responsibility? Environ Sci Pol 48:89–98. https://doi.org/10.1016/j.envsci.2014.12.011
Nath PK, Behera B (2011) A critical review of impact of and adaptation to climate change in developed and developing economies. Environ Dev Sustain 13(1):141–162. https://doi.org/10.1007/s10668-010-9253-9
Nay JJ, Abkowitz M, Chu E, Gallagher D, Wright H (2014) A review of decision-support models for adaptation to climate change in the context of development. Clim Dev 6 (4):357–367. https://doi.org/10.1080/17565529.2014.912196
Nightingale AJ (2017) Power and politics in climate change adaptation efforts: struggles over authority and recognition in the context of political instability. Geoforum 84:11–20. https://doi.org/10.1016/j.geoforum.2017.05.011
Njoku JD (2006) Analysis of the effect of global warming on forest of southeastern Nigeria using remotely sensed data. A PhD thesis Imo State University, Owerri, Nigeria
Nwagbara MO, Chima GN, Ndukwe-Okoye N (2013) Climate change and soil erosion in the derived savanna of southeastern Nigeria. Being a paper presented at the 37th annual conference of the soil science society of Igeria held at Lafia, Nasara State, 11–15 March
Nzeadibe T, Egbule C, Chukwuone N, Agu V (2011) Climate change awareness and adaptation in the Niger Delta Rgion of Nigeria. African Technology Policy Studies Network, Nairobi
Ogundiya IS (2010) Democracy and good governance: Nigeria's dilemma. Afr J Polit Sci Int Rel 4 (6):201–208
Olajide BE, Quadri MO, Ojakorotu V (2018) Climate change, human security and good governance in Nigeria. Afr Renaiss 15(3):173–196
Orji S (2018) It rains, it pours, it floods: Nigeria's growing seasonal problem. African Argument. Retrieved from https://africanarguments.org/2018/11/15/nigeria-floods-growing-problem/
Otto IM, Reckien D, Reyer CPO, Marcus R, Le Masson V, Jones L, . . . Serdeczny O (2017) Social vulnerability to climate change: a review of concepts and evidence. Reg Environ Chang 17 (6):1651–1662. https://doi.org/10.1007/s10113-017-1105-9.
Oulu M (2015) Climate change governance: emerging legal and institutional frameworks for developing countries. In: Leal Filho W (ed) Handbook of climate change adaptation. Springer Berlin Heidelberg, Berlin/Heidelberg, pp 1–20
Paavola J, Adger WN (2006) Fair adaptation to climate change. Ecol Econ 56(4):594–609. https://doi.org/10.1016/j.ecolecon.2005.03.015
Rauken T, Mydske PK, Winsvold M (2015) Mainstreaming climate change adaptation at the local level. Local Environment 20(4):408–423. https://doi.org/10.1080/13549839.2014.880412
Rawls J (1971) A theory of justice, Revised edn. The Belknap Press of Harvard Press, Cambridge, MA
Rübbelke DTG (2011) International support of climate change policies in developing countries: strategic, moral and fairness aspects. Ecol Econ 70(8):1470–1480. https://doi.org/10.1016/j.ecolecon.2011.03.007
Saraswat C, Kumar P (2016) Climate justice in lieu of climate change: a sustainable approach to respond to the climate change injustice and an awakening of the environmental movement. Energy Ecol Environ 1(2):67–74. https://doi.org/10.1007/s40974-015-0001-8

Schlosberg D (2013) Theorising environmental justice: the expanding sphere of a discourse. Environ Polit 22(1):37–55. https://doi.org/10.1080/09644016.2013.755387

Simonsson L, Åsa Gerger S, Karin A, Oskar W, Richard J. T. Klein (2011) Perceptions of risk and limits to climate change adaptation: case studies of two Swedish urban regions. In: Ford JD, Berrang-Ford L (eds) Climate change adaptation in developed nations: from theory to practice. Dordrecht: Springer Netherlands, pp 321–334

Stallworthy M (2009) Environmental justice imperatives for an era of climate change. J Law Soc 36(1):55–74

Tabbo AM, Amadou Z (2017) Assessing newly introduced climate change adaptation strategy packages among rural households: evidence from Kaou local government area, Tahoua State, Niger Republic. Jamba (Potchefstroom, South Africa) 9(1):383–383. https://doi.org/10.4102/jamba.v9i1.383

Thomas DSG, Twyman C (2005) Equity and justice in climate change adaptation amongst natural-resource dependent societies. Glob Environ Chang 15:115–124

Urwin K, Jordan A (2008) Does public policy support or undermine climate change adaptation? Exploring policy interplay across different scales of governance. Glob Environ Chang 18:180–191

Vogel B, Henstra D (2015) Studying local climate adaptation: a heuristic research framework for comparative policy analysis. Glob Environ Chang 31:110–120. https://doi.org/10.1016/j.gloenvcha.2015.01.001

Vogel C, Moser SC, Kasperson RE, Dabelko GD (2007) Linking vulnerability, adaptation, and resilience science to practice: pathways, players, and partnerships. Glob Environ Chang 17 (3):349–364. https://doi.org/10.1016/j.gloenvcha.2007.05.002

Wisner B, Blaikie P, Cannon T, Davis I (2004) At risk: natural hazards, people's vulnerability and disasters, 2nd edn. Routledge, London

Permissions

All chapters in this book were first published by Springer; hereby published with permission under the Creative Commons Attribution License or equivalent. Every chapter published in this book has been scrutinized by our experts. Their significance has been extensively debated. The topics covered herein carry significant findings which will fuel the growth of the discipline. They may even be implemented as practical applications or may be referred to as a beginning point for another development.

The contributors of this book come from diverse backgrounds, making this book a truly international effort. This book will bring forth new frontiers with its revolutionizing research information and detailed analysis of the nascent developments around the world.

We would like to thank all the contributing authors for lending their expertise to make the book truly unique. They have played a crucial role in the development of this book. Without their invaluable contributions this book wouldn't have been possible. They have made vital efforts to compile up to date information on the varied aspects of this subject to make this book a valuable addition to the collection of many professionals and students.

This book was conceptualized with the vision of imparting up-to-date information and advanced data in this field. To ensure the same, a matchless editorial board was set up. Every individual on the board went through rigorous rounds of assessment to prove their worth. After which they invested a large part of their time researching and compiling the most relevant data for our readers.

The editorial board has been involved in producing this book since its inception. They have spent rigorous hours researching and exploring the diverse topics which have resulted in the successful publishing of this book. They have passed on their knowledge of decades through this book. To expedite this challenging task, the publisher supported the team at every step. A small team of assistant editors was also appointed to further simplify the editing procedure and attain best results for the readers.

Apart from the editorial board, the designing team has also invested a significant amount of their time in understanding the subject and creating the most relevant covers. They scrutinized every image to scout for the most suitable representation of the subject and create an appropriate cover for the book.

The publishing team has been an ardent support to the editorial, designing and production team. Their endless efforts to recruit the best for this project, has resulted in the accomplishment of this book. They are a veteran in the field of academics and their pool of knowledge is as vast as their experience in printing. Their expertise and guidance has proved useful at every step. Their uncompromising quality standards have made this book an exceptional effort. Their encouragement from time to time has been an inspiration for everyone.

The publisher and the editorial board hope that this book will prove to be a valuable piece of knowledge for researchers, students, practitioners and scholars across the globe.

LIST OF CONTRIBUTORS

Ayansina Ayanlade, Isaac Ayo Oluwatimilehin and Godwin Atai
Department of Geography, Obafemi Awolowo University, Ile-Ife, Nigeria

Adeola A. Oladimeji
Department of Microbiology, University of Ibadan, Ibadan, Nigeria

Damilola T. Agbalajobi
Department of Political Science, Obafemi Awolowo University, Ile-Ife, Nigeria

Hilda Manzi and Joseph P. Gweyi-Onyango
Department of Agricultural Science and Technology, Kenyatta University, Nairobi, Kenya

Oseni Taiwo Amoo
Risk and Vulnerabilty Science Centre, Walter Sasilu University, Eastern Cape, South Africa

Hammed Olabode Ojugbele
Regional and Local Economic Development Initiative, University of KwaZulu-Natal, Westville, South Africa

Abdultaofeek Abayomi
Department of Information and Communication Technology, Mangosuthu University of Technology, Umlazi, Durban, South Africa

Pushpendra Kumar Singh
Water Resources Systems Division, National Institute of Hydrology, Roorkee, India

Philip Olanrewaju Eniola
Department of Agricultural Technology, The Oke-Ogun Polytechnic, Saki, Oyo State, Nigeria

Michael Robert Nkuba, Raban Chanda, Gagoitseope Mmopelwa and David Lesolle
Department of Environmental sciences, Faculty of Science University of Botswana, Gaborone, Botswana

Akintayo Adedoyin
Department of Physics, Faculty of Science, University of Botswana, Gaborone, Botswana

Margaret Najjingo Mangheni
Department of Extension and Innovation Studies, College of Agricultural and Environmental Sciences, Makerere University Kampala, Kampala, Uganda

Edward Kato
International Food Policy and Research Institute, Washington, DC, USA

Erimma Gloria Orie
Department of Private and Property Law, National Open University of Nigeria, Abuja, Nigeria

Peter Urich, Yinpeng Li and Sennye Masike
International Global Change Institute and CLIMsystems Ltd, Hamilton, New Zealand

Daniel M. Nzengya
Department of Social Sciences, St Paul's University, Limuru, Kenya

John Kibe Maguta
Faculty of Social Science, St Paul's University, Limuru, Kenya

V. A. Tanimonure
Agricultural Economics Department, Obafemi Awolowo University, Ile-Ife, Nigeria

John Saviour Yaw Eleblu and Eric Yirenkyi Danquah
West Africa Centre for Crop Improvement, University of Ghana, College of Basic and Applied Sciences, Accra, Ghana

Eugene Tenkorang Darko
Geography and Resource Development, University of Ghana, Legon, Ghana

Chinwe Philomina Oramah and Odd Einar Olsen
Department of Safety, Economics, and Planning, University of Stavanger, Stavanger, Norway

Index

Printed in the USA
CPSIA information can be obtained
at www.ICGtesting.com
JSHW051350091023
49903JS00006B/90